www.EffortlessMath.com

... So Much More Online!

✓ FREE Math lessons

✓ More Math learning books!

✓ Mathematics Worksheets

✓ Online Math Tutors

Need a PDF version of this book?

Visit www.EffortlessMath.com

Or send email to: info@EffortlessMath.com

ATI TEAS 6

Mathematics

Prep 2019

A Comprehensive Review and Ultimate Guide to the ATI TEAS 6 Math Test

By

Reza Nazari & Ava Ross

All inquiries should be addressed to:

info@effortlessMath.com

www.EffortlessMath.com

ISBN-13: 978-1-970036-07-7

ISBN-10: 1-970036-07-9

Published by: Effortless Math Education

www.EffortlessMath.com

Description

ATI TEAS 6 Mathematics Prep 2019 provides students with the confidence and math skills they need to succeed on the ATI TEAS 6 Math, building a solid foundation of basic Math topics with abundant exercises for each topic. It is designed to address the needs of TEAS 6 test takers who must have a working knowledge of basic Math.

This comprehensive book with over 2,500 sample questions and 2 complete ATI TEAS 6 tests is all you need to fully prepare for the TEAS 6 Math. It will help you learn everything you need to ace the math section of the TEAS 6.

Effortless Math unique study program provides you with an in-depth focus on the math portion of the exam, helping you master the math skills that students find the most troublesome.
This book contains most common sample questions that are most likely to appear in the mathematics section of the ATI TEAS 6.

Inside the pages of this comprehensive ATI TEAS 6 Math book, students can learn basic math operations in a structured manner with a complete study program to help them understand essential math skills. It also has many exciting features, including:

- Dynamic design and easy-to-follow activities
- A fun, interactive and concrete learning process
- Targeted, skill-building practices
- Fun exercises that build confidence
- Math topics are grouped by category, so you can focus on the topics you struggle on
- All solutions for the exercises are included, so you will always find the answers
- 2 Complete ATI TEAS 6 Math Practice Tests that reflect the format and question types on ATI TEAS 6

ATI TEAS 6 Mathematics Prep 2019 is an incredibly useful tool for those who want to review all topics being covered on the ATI TEAS 6 test. It efficiently and effectively reinforces learning outcomes through engaging questions and repeated practice, helping you to quickly master basic Math skills.

About the Author

Reza Nazari is the author of more than 100 Math learning books including:
– **Math and Critical Thinking Challenges:** For the Middle and High School Student
– **GED Math in 30 Days**
– **ASVAB Math Workbook 2018 - 2019**
– **Effortless Math Education Workbooks**
– **and many more Mathematics books ...**

Reza is also an experienced Math instructor and a test–prep expert who has been tutoring students since 2008. Reza is the founder of Effortless Math Education, a tutoring company that has helped many students raise their standardized test scores—and attend the colleges of their dreams. Reza provides an individualized custom learning plan and the personalized attention that makes a difference in how students view math.

You can contact Reza via email at:
reza@EffortlessMath.com

Find Reza's professional profile at:
goo.gl/zoC9rJ

Contents

Chapter 1: Arithmetic

Topics that you'll learn in this chapter:

- ✓ Simplifying Fractions
- ✓ Adding and Subtracting Fractions
- ✓ Multiplying and Dividing Fractions
- ✓ Adding Mixed Numbers
- ✓ Subtract Mixed Numbers
- ✓ Multiplying Mixed Numbers
- ✓ Dividing Mixed Numbers
- ✓ Comparing Decimals
- ✓ Rounding Decimals
- ✓ Adding and Subtracting Decimals
- ✓ Multiplying and Dividing Decimals
- ✓ Converting Between Fractions, Decimals and Mixed Numbers
- ✓ Factoring Numbers
- ✓ Greatest Common Factor
- ✓ Least Common Multiple
- ✓ Divisibility Rules

Simplifying Fractions

Helpful		Example:
Hints	– Evenly divide both the top and bottom of the fraction by 2, 3, 5, 7, ... etc. – Continue until you can't go any further.	$\frac{4}{12} = \frac{2}{6} = \frac{1}{3}$

✍ *Simplify the fractions.*

1) $\frac{22}{36}$

2) $\frac{8}{10}$

3) $\frac{12}{18}$

4) $\frac{6}{8}$

5) $\frac{13}{39}$

6) $\frac{5}{20}$

7) $\frac{16}{36}$

8) $\frac{18}{36}$

9) $\frac{20}{50}$

10) $\frac{6}{54}$

11) $\frac{45}{81}$

12) $\frac{21}{28}$

13) $\frac{35}{56}$

14) $\frac{52}{64}$

15) $\frac{13}{65}$

16) $\frac{44}{77}$

17) $\frac{21}{42}$

18) $\frac{15}{36}$

19) $\frac{9}{24}$

20) $\frac{20}{80}$

21) $\frac{25}{45}$

Adding and Subtracting Fractions

Helpful *Hints*	– For "like" fractions (fractions with the same denominator), add or subtract the numerators and write the answer over the common denominator.

– Find equivalent fractions with the same denominator before you can add or subtract fractions with different denominators.

– Adding and Subtracting with the same denominator:

$$\frac{a}{b} + \frac{c}{b} = \frac{a+c}{b}$$
$$\frac{a}{b} - \frac{c}{b} = \frac{a-c}{b}$$

– Adding and Subtracting fractions with different denominators:

$$\frac{a}{b} + \frac{c}{d} = \frac{ad+cb}{bd}$$
$$\frac{a}{b} - \frac{c}{d} = \frac{ad-cb}{bd}$$

✍ Add fractions.

1) $\frac{2}{3} + \frac{1}{2}$

2) $\frac{3}{5} + \frac{1}{3}$

3) $\frac{5}{6} + \frac{1}{2}$

4) $\frac{7}{4} + \frac{5}{9}$

5) $\frac{2}{5} + \frac{1}{5}$

6) $\frac{3}{7} + \frac{1}{2}$

7) $\frac{3}{4} + \frac{2}{5}$

8) $\frac{2}{3} + \frac{1}{5}$

9) $\frac{16}{25} + \frac{3}{5}$

✍ Subtract fractions.

10) $\frac{4}{5} - \frac{2}{5}$

11) $\frac{3}{5} - \frac{2}{7}$

12) $\frac{1}{2} - \frac{1}{3}$

13) $\frac{8}{9} - \frac{3}{5}$

14) $\frac{3}{7} - \frac{3}{14}$

15) $\frac{4}{15} - \frac{1}{10}$

16) $\frac{3}{4} - \frac{13}{18}$

17) $\frac{5}{8} - \frac{2}{5}$

18) $\frac{1}{2} - \frac{1}{9}$

Multiplying and Dividing Fractions

Helpful Hints	– **Multiplying fractions:** multiply the top numbers and multiply the bottom numbers. – **Dividing fractions:** Keep, Change, Flip Keep first fraction, change division sign to multiplication, and flip the numerator and denominator of the second fraction. Then, solve!	**Example:** $$\frac{a}{b} \times \frac{c}{d} = \frac{a \times c}{b \times d}$$ $$\frac{a}{b} \div \frac{c}{d} = \frac{a}{b} \times \frac{d}{c} = \frac{ad}{bc}$$

✎ *Multiplying fractions. Then simplify.*

1) $\dfrac{1}{5} \times \dfrac{2}{3}$

2) $\dfrac{3}{4} \times \dfrac{2}{3}$

3) $\dfrac{2}{5} \times \dfrac{3}{7}$

4) $\dfrac{3}{8} \times \dfrac{1}{3}$

5) $\dfrac{3}{5} \times \dfrac{2}{5}$

6) $\dfrac{7}{9} \times \dfrac{1}{3}$

7) $\dfrac{2}{3} \times \dfrac{3}{8}$

8) $\dfrac{1}{4} \times \dfrac{1}{3}$

9) $\dfrac{5}{7} \times \dfrac{7}{12}$

✎ *Dividing fractions.*

10) $\dfrac{2}{9} \div \dfrac{1}{4}$

11) $\dfrac{1}{2} \div \dfrac{1}{3}$

12) $\dfrac{6}{11} \div \dfrac{3}{4}$

13) $\dfrac{11}{14} \div \dfrac{1}{10}$

14) $\dfrac{3}{5} \div \dfrac{5}{9}$

15) $\dfrac{1}{2} \div \dfrac{1}{2}$

16) $\dfrac{3}{5} \div \dfrac{1}{5}$

17) $\dfrac{12}{21} \div \dfrac{3}{7}$

18) $\dfrac{5}{14} \div \dfrac{9}{10}$

Adding Mixed Numbers

Helpful *Hints*	Use the following steps for both adding and subtracting mixed numbers. – Find the Least Common Denominator (LCD) – Find the equivalent fractions for each mixed number. – Add fractions after finding common denominator. – Write your answer in lowest terms.	**Example:** $1\frac{3}{4} + 2\frac{3}{8} = 4\frac{1}{8}$

✏️*Add.*

1) $4\frac{1}{2} + 5\frac{1}{2}$

2) $2\frac{3}{8} + 3\frac{1}{8}$

3) $6\frac{1}{5} + 3\frac{2}{5}$

4) $1\frac{1}{3} + 2\frac{2}{3}$

5) $5\frac{1}{6} + 5\frac{1}{2}$

6) $3\frac{1}{3} + 1\frac{1}{3}$

7) $1\frac{10}{11} + 1\frac{1}{3}$

8) $2\frac{3}{6} + 1\frac{1}{2}$

9) $5\frac{3}{5} + 5\frac{1}{5}$

10) $7 + \frac{1}{5}$

11) $1\frac{5}{7} + \frac{1}{3}$

12) $2\frac{1}{4} + 1\frac{1}{2}$

Subtract Mixed Numbers

Helpful *Hints*	Use the following steps for both adding and subtracting mixed numbers. Find the Least Common Denominator (LCD) − Find the equivalent fractions for each mixed number. − Add or subtract fractions after finding common denominator. − Write your answer in lowest terms.	**Example:** $5\frac{2}{3} - 3\frac{2}{7} = 2\frac{8}{21}$

✎ *Subtract.*

1) $4\frac{1}{2} - 3\frac{1}{2}$

2) $3\frac{3}{8} - 3\frac{1}{8}$

3) $6\frac{3}{5} - 5\frac{1}{5}$

4) $2\frac{1}{3} - 1\frac{2}{3}$

5) $6\frac{1}{6} - 5\frac{1}{2}$

6) $3\frac{1}{3} - 1\frac{1}{3}$

7) $2\frac{10}{11} - 1\frac{1}{3}$

8) $2\frac{1}{2} - 1\frac{1}{2}$

9) $6\frac{3}{5} - 2\frac{1}{5}$

10) $7\frac{2}{5} - 1\frac{1}{5}$

11) $2\frac{5}{7} - 1\frac{1}{3}$

12) $2\frac{1}{4} - 1\frac{1}{2}$

Multiplying Mixed Numbers

Helpful	1- Convert the mixed numbers to improper fractions.	**Example:**
Hints	2- Multiply fractions and simplify if necessary.	$2\frac{1}{3} \times 5\frac{3}{7} =$
	$a\frac{c}{b} = a + \frac{c}{b} = \frac{ab+c}{b}$	$\frac{7}{3} \times \frac{38}{7} = \frac{38}{3} = 12\frac{2}{3}$

✎ *Find each product.*

1) $1\frac{2}{3} \times 1\frac{1}{4}$

2) $1\frac{3}{5} \times 1\frac{2}{3}$

3) $1\frac{2}{3} \times 3\frac{2}{7}$

4) $4\frac{1}{8} \times 1\frac{2}{5}$

5) $2\frac{2}{5} \times 3\frac{1}{5}$

6) $1\frac{1}{3} \times 1\frac{2}{3}$

7) $1\frac{5}{8} \times 2\frac{1}{2}$

8) $3\frac{2}{5} \times 2\frac{1}{5}$

9) $2\frac{2}{3} \times 4\frac{1}{4}$

10) $2\frac{3}{5} \times 1\frac{2}{4}$

11) $1\frac{1}{3} \times 1\frac{1}{4}$

12) $3\frac{2}{5} \times 1\frac{1}{5}$

Dividing Mixed Numbers

Helpful *Hints*	1- Convert the mixed numbers to improper fractions. 2- Divide fractions and simplify if necessary. $a\dfrac{c}{b} = a + \dfrac{c}{b} = \dfrac{ab+c}{b}$	**Example:** $2\dfrac{1}{3} \div 5\dfrac{3}{7} =$ $\dfrac{7}{3} \div \dfrac{38}{7} = \dfrac{7}{3} \times \dfrac{7}{38} = \dfrac{49}{108}$

✎ *Find each quotient.*

1) $2\dfrac{1}{5} \div 2\dfrac{1}{2}$

5) $4\dfrac{1}{8} \div 2\dfrac{2}{4}$

9) $3\dfrac{1}{5} \div 1\dfrac{1}{2}$

2) $2\dfrac{3}{5} \div 1\dfrac{1}{3}$

6) $3\dfrac{1}{2} \div 2\dfrac{3}{5}$

10) $4\dfrac{3}{5} \div 2\dfrac{1}{3}$

3) $3\dfrac{1}{6} \div 4\dfrac{2}{3}$

7) $3\dfrac{5}{9} \div 1\dfrac{2}{5}$

11) $6\dfrac{1}{6} \div 1\dfrac{2}{3}$

4) $1\dfrac{2}{3} \div 3\dfrac{1}{3}$

8) $2\dfrac{2}{7} \div 1\dfrac{1}{2}$

12) $2\dfrac{2}{3} \div 1\dfrac{1}{3}$

Comparing Decimals

Helpful	-	**Decimals:** is a fraction written in a special form. For example, instead of writing $\frac{1}{2}$ you can write 0.5.	**Example:**
Hints	-	**For comparing:** Equal to = Less than < Greater than > Greater than or equal ≥ Less than or equal ≤	2.67 > 0.267

✍ *Write the correct comparison symbol (>, < or =).*

1) 1.25   2.3

2) 0.5   0.23

3) 3.2   3.2

4) 4.58   45.8

5) 2.75   0.275

6) 5.2   5

7) 3.1   0.31

8) 6.33   0.733

9) 8   0.8

10) 4.56   0.456

11) 1.12   1.14

12) 2.77   2.78

13) 6.08   6.11

14) 1.11   0.211

15) 2.6   2.55

16) 1.24   1.25

17) 5.52   0.552

18) 0.33   0.033

19) 14.4   14.4

20) 0.05   0.50

21) 0.59   0.7

22) 0.5   0.05

23) 0.90   0.9

24) 0.27   0.4

Rounding Decimals

Helpful

Hints

We can round decimals to a certain accuracy or number of decimal places. This is used to make calculation easier to do and results easier to understand, when exact values are not too important.

First, you'll need to remember your place values:

12.4567

1: tens	2: ones	4: tenths

5: hundredths	6: thousandths	7: ten thousandths

Example:

6.37 = 6

✍ *Round each decimal number to the nearest place indicated.*

1) 0.2<u>3</u>	9) 1.6<u>2</u>9	17) 70.7<u>8</u>
2) 4.<u>0</u>4	10) 6.<u>3</u>959	18) 61<u>5</u>.755
3) 5.<u>6</u>23	11) <u>1</u>.9	19) 1<u>6</u>.4
4) 0.<u>2</u>66	12) <u>5</u>.2167	20) 9<u>5</u>.81
5) <u>6</u>.37	13) 5.<u>8</u>63	21) <u>2</u>.408
6) 0.8<u>8</u>	14) 8.<u>5</u>4	22) 7<u>6</u>.3
7) 8.<u>2</u>4	15) 80.<u>6</u>9	23) 116.<u>5</u>14
8) <u>7</u>.0760	16) 6<u>5</u>.85	24) 8.<u>0</u>6

Adding and Subtracting Decimals

Helpful *Hints*	1– Line up the numbers. 2– Add zeros to have same number of digits for both numbers. 3– Add or Subtract using column addition or subtraction.	**Example:** $$\begin{array}{r} 16.18 \\ -\ 13.45 \\ \hline 2.73 \end{array}$$

✎ **Add and subtract decimals.**

1) $\begin{array}{r} 15.14 \\ -\ 12.18 \\ \hline \end{array}$

3) $\begin{array}{r} 82.56 \\ +\ 12.28 \\ \hline \end{array}$

5) $\begin{array}{r} 90.37 \\ +\ 56.97 \\ \hline \end{array}$

2) $\begin{array}{r} 65.72 \\ +\ 43.67 \\ \hline \end{array}$

4) $\begin{array}{r} 34.18 \\ -\ 23.45 \\ \hline \end{array}$

6) $\begin{array}{r} 45.78 \\ -\ 23.39 \\ \hline \end{array}$

✎ **Solve.**

7) ____ $+ 1.3 = 4.8$

8) $4.2 +$ ____ $= 11.6$

9) $9.9 +$ ____ $= 16$

10) $6.9 +$ ____ $= 16.4$

11) ____ $+ 5.1 = 8.6$

12) ____ $+ 7.9 = 15.2$

Multiplying and Dividing Decimals

Helpful *Hints*	**For Multiplication:**
	– Set up and multiply the numbers as you do with whole numbers.
	– Count the total number of decimal places in both of the factors.
	– Place the decimal point in the product.
	For Division:
	– If the divisor is not a whole number, move decimal point to right to make it a whole number. Do the same for dividend.
	– Divide similar to whole numbers.

✎ Find each product.

1)
$$\begin{array}{r} 4.5 \\ \times\ 1.6 \\ \hline \end{array}$$

4)
$$\begin{array}{r} 8.9 \\ \times\ 9.7 \\ \hline \end{array}$$

7)
$$\begin{array}{r} 5.7 \\ \times\ 7.8 \\ \hline \end{array}$$

2)
$$\begin{array}{r} 7.7 \\ \times\ 9.9 \\ \hline \end{array}$$

5)
$$\begin{array}{r} 15.1 \\ \times\ 12.6 \\ \hline \end{array}$$

8)
$$\begin{array}{r} 98.20 \\ \times\ 100 \\ \hline \end{array}$$

3)
$$\begin{array}{r} 2.6 \\ \times\ 1.5 \\ \hline \end{array}$$

6)
$$\begin{array}{r} 6.9 \\ \times\ 3.3 \\ \hline \end{array}$$

9)
$$\begin{array}{r} 23.99 \\ \times\ 1000 \\ \hline \end{array}$$

✎ Find each quotient.

10) $9.2 \div 3.6$

11) $27.6 \div 3.8$

12) $12.6 \div 4.7$

13) $6.5 \div 8.1$

14) $1.4 \div 10$

15) $3.6 \div 100$

16) $4.24 \div 10$

17) $14.6 \div 100$

18) $1.8 \div 1000$

Converting Between Fractions, Decimals and Mixed Numbers

Helpful *Hints*	**Fraction to Decimal:** – Divide the top number by the bottom number. **Decimal to Fraction:** – Write decimal over 1. – Multiply both top and bottom by 10 for every digit on the right side of the decimal point. – Simplify.

✎ *Convert fractions to decimals.*

1) $\dfrac{9}{10}$ 4) $\dfrac{2}{5}$ 7) $\dfrac{12}{10}$

2) $\dfrac{56}{100}$ 5) $\dfrac{3}{9}$ 8) $\dfrac{8}{5}$

3) $\dfrac{3}{4}$ 6) $\dfrac{40}{50}$ 9) $\dfrac{69}{10}$

✎ *Convert decimal into fraction or mixed numbers.*

10) 0.3   14) 0.8   18) 0.08

11) 4.5   15) 0.25   19) 0.45

12) 2.5   16) 0.14   20) 2.6

13) 2.3   17) 0.2   21) 5.2

Factoring Numbers

Helpful *Hints*	- Factoring numbers means to break the numbers into their prime factors. - First few prime numbers: 2, 3, 5, 7, 11, 13, 17, 19	**Example:** $12 = 2 \times 2 \times 3$

✎ List all positive factors of each number.

1) 68   6) 78   11) 54

2) 56   7) 50   12) 28

3) 24   8) 98   13) 55

4) 40   9) 45   14) 85

5) 86   10) 26   15) 48

✎ List the prime factorization for each number.

16) 50   19) 21   22) 26

17) 25   20) 45   23) 86

18) 69   21) 68   24) 93

Greatest Common Factor

Helpful	- List the prime factors of each number.	**Example:**
Hints	- Multiply common prime factors.	$200 = 2 \times 2 \times 2 \times 5 \times 5$
		$60 = 2 \times 2 \times 3 \times 5$
		GCF $(200, 60) = 2 \times 2 \times 5 = 20$

✎ _Find the GCF for each number pair._

1) 20, 30

2) 4, 14

3) 5, 45

4) 68, 12

5) 5, 12

6) 15, 27

7) 3, 24

8) 34, 6

9) 4, 10

10) 5, 3

11) 6, 16

12) 30, 3

13) 24, 28

14) 70, 10

15) 45, 8

16) 90, 35

17) 78, 34

18) 55, 75

19) 60, 72

20) 100, 78

21) 30, 40

Least Common Multiple

Helpful	- Find the GCF for the two numbers. - Divide that GCF into either number. - Take that answer and multiply it by the other number.	**Example:** LCM (200, 60): GCF is 20 $200 \div 20 = 10$ $10 \times 60 = 600$
Hints		

✎ *Find the LCM for each number pair.*

1) 4, 14

2) 5, 15

3) 16, 10

4) 4, 34

5) 8, 3

6) 12, 24

7) 9, 18

8) 5, 6

9) 8, 19

10) 9, 21

11) 19, 29

12) 7, 6

13) 25, 6

14) 4, 8

15) 30, 10, 50

16) 18, 36, 27

17) 12, 8, 18

18) 8, 18, 4

19) 26, 20, 30

20) 10, 4, 24

21) 15, 30, 45

Divisibility Rules

Helpful	-	Divisibility means that a number can be divided by other numbers evenly.	**Example:** 24 is divisible by 6, because 24 ÷ 6 = 4
Hints			

✎ *Use the divisibility rules to find the factors of each number.*

8   2 3 4 5 6 7 8 9 10

1) 16   2 3 4 5 6 7 8 9 10

2) 10   2 3 4 5 6 7 8 9 10

3) 15   2 3 4 5 6 7 8 9 10

4) 28   2 3 4 5 6 7 8 9 10

5) 36   2 3 4 5 6 7 8 9 10

6) 15   2 3 4 5 6 7 8 9 10

7) 27   2 3 4 5 6 7 8 9 10

8) 70   2 3 4 5 6 7 8 9 10

9) 57   2 3 4 5 6 7 8 9 10

10) 102   2 3 4 5 6 7 8 9 10

11) 144   2 3 4 5 6 7 8 9 10

12) 75   2 3 4 5 6 7 8 9 10

Answers of Worksheets – Chapter 1

Simplifying Fractions

1) $\frac{11}{18}$

2) $\frac{4}{5}$

3) $\frac{2}{3}$

4) $\frac{3}{4}$

5) $\frac{1}{3}$

6) $\frac{1}{4}$

7) $\frac{4}{9}$

8) $\frac{1}{2}$

9) $\frac{2}{5}$

10) $\frac{1}{9}$

11) $\frac{5}{9}$

12) $\frac{3}{4}$

13) $\frac{5}{8}$

14) $\frac{13}{16}$

15) $\frac{1}{5}$

16) $\frac{4}{7}$

17) $\frac{1}{2}$

18) $\frac{5}{12}$

19) $\frac{3}{8}$

20) $\frac{1}{4}$

21) $\frac{5}{9}$

Adding and Subtracting Fractions

1) $\frac{7}{6}$

2) $\frac{14}{15}$

3) $\frac{4}{3}$

4) $\frac{83}{36}$

5) $\frac{3}{5}$

6) $\frac{13}{14}$

7) $\frac{23}{20}$

8) $\frac{13}{15}$

9) $\frac{31}{25}$

10) $\frac{2}{5}$

11) $\frac{11}{35}$

12) $\frac{1}{6}$

13) $\frac{13}{45}$

14) $\frac{3}{14}$

15) $\frac{1}{6}$

16) $\frac{1}{36}$

17) $\frac{9}{40}$

18) $\frac{7}{18}$

Multiplying and Dividing Fractions

1) $\dfrac{2}{15}$

2) $\dfrac{1}{2}$

3) $\dfrac{6}{35}$

4) $\dfrac{1}{8}$

5) $\dfrac{6}{25}$

6) $\dfrac{7}{27}$

7) $\dfrac{1}{4}$

8) $\dfrac{1}{12}$

9) $\dfrac{5}{12}$

10) $\dfrac{8}{9}$

11) $\dfrac{3}{2}$

12) $\dfrac{8}{11}$

13) $\dfrac{55}{7}$

14) $\dfrac{27}{25}$

15) 1

16) 3

17) $\dfrac{4}{3}$

18) $\dfrac{25}{63}$

Adding Mixed Numbers

1) 10

2) $5\dfrac{1}{2}$

3) $9\dfrac{3}{5}$

4) 4

5) $10\dfrac{2}{3}$

6) $4\dfrac{2}{3}$

7) $3\dfrac{8}{33}$

8) 4

9) $10\dfrac{4}{5}$

10) $7\dfrac{1}{5}$

11) $2\dfrac{1}{21}$

12) $3\dfrac{3}{4}$

Subtract Mixed Numbers

1) 1

2) $\dfrac{1}{4}$

3) $1\dfrac{2}{5}$

4) $\dfrac{2}{3}$

5) $\dfrac{2}{3}$

6) 2

7) $1\dfrac{19}{33}$

8) 1

9) $4\dfrac{2}{5}$

10) $6\dfrac{1}{5}$

11) $1\dfrac{8}{21}$

12) $\dfrac{3}{4}$

Multiplying Mixed Numbers

1) $2\frac{1}{12}$

2) $2\frac{2}{3}$

3) $5\frac{10}{21}$

4) $5\frac{31}{40}$

5) $7\frac{17}{25}$

6) $2\frac{2}{9}$

7) $4\frac{1}{16}$

8) $7\frac{12}{25}$

9) $11\frac{1}{3}$

10) $3\frac{9}{10}$

11) $1\frac{2}{3}$

12) $4\frac{2}{25}$

Dividing Mixed Numbers

1) $\frac{22}{25}$

2) $1\frac{19}{20}$

3) $\frac{19}{28}$

4) $\frac{1}{2}$

5) $1\frac{13}{20}$

6) $1\frac{9}{26}$

7) $2\frac{34}{63}$

8) $1\frac{11}{21}$

9) $2\frac{2}{15}$

10) $1\frac{34}{35}$

11) $3\frac{7}{10}$

12) 2

Comparing Decimals

1) $1.25 < 2.3$

2) $0.5 > 0.23$

3) $3.2 = 3.2$

4) $4.58 < 45.8$

5) $2.75 > 0.275$

6) $5.2 > 5$

7) $3.1 > 0.31$

8) $6.33 > 0.733$

9) $8 > 0.8$

10) $4.56 > 0.456$

11) $1.12 < 1.14$

12) $2.77 < 2.78$

13) $6.08 < 6.11$

14) $1.11 > 0.211$

15) $2.6 > 2.55$

16) $1.24 < 1.25$

17) $5.52 > 0.552$

18) $0.33 > 0.033$

19) $14.4 = 14.4$

20) $0.05 < 0.50$

21) $0.59 < 0.7$

22) $0.5 > 0.05$

23) $0.90 = 0.9$

24) $0.27 < 0.4$

Rounding Decimals

1) 0.2
2) 4.0
3) 5.6
4) 0.3
5) 6
6) 0.9
7) 8.2
8) 7

9) 1.63
10) 6.4
11) 2
12) 5
13) 5.9
14) 8.5
15) 81
16) 66

17) 70.8
18) 616
19) 16
20) 96
21) 2
22) 76
23) 116.5
24) 8.1

Adding and Subtracting Decimals

1) 2.96
2) 109.39
3) 94.84
4) 10.73

5) 147.34
6) 22.39
7) 3.5
8) 7.4

9) 6.1
10) 9.5
11) 3.5
12) 7.3

Multiplying and Dividing Decimals

1) 7.2
2) 76.23
3) 3.9
4) 86.33
5) 190.26
6) 22.77

7) 44.46
8) 9820
9) 23990
10) 2.5555…
11) 7.2631…
12) 2.6808…

13) 0.8024…
14) 0.14
15) 0.036
16) 0.424
17) 0.146
18) 0.0018

Converting Between Fractions, Decimals and Mixed Numbers

1) 0.9
2) 0.56
3) 0.75
4) 0.4
5) 0.333…
6) 0.8

7) 1.2
8) 1.6
9) 6.9
10) $\frac{3}{10}$
11) $4\frac{1}{2}$

12) $2\frac{1}{2}$
13) $2\frac{3}{10}$
14) $\frac{4}{5}$
15) $\frac{1}{4}$

16) $\dfrac{7}{50}$ 18) $\dfrac{2}{25}$ 20) $2\dfrac{3}{5}$

17) $\dfrac{1}{5}$ 19) $\dfrac{9}{20}$ 21) $5\dfrac{1}{5}$

Factoring Numbers

1) 1, 2, 4, 17, 34, 68
2) 1, 2, 4, 7, 8, 14, 28, 56
3) 1, 2, 3, 4, 6, 8, 12, 24
4) 1, 2, 4, 5, 8, 10, 20, 40
5) 1, 2, 43, 86
6) 1, 2, 3, 6, 13, 26, 39, 78
7) 1, 2, 5, 10, 25, 50
8) 1, 2, 7, 14, 49, 98
9) 1, 3, 5, 9, 15, 45
10) 1, 2, 13, 26
11) 1, 2, 3, 6, 9, 18, 27, 54
12) 1, 2, 4, 7, 14, 28

13) 1, 5, 11, 55
14) 1, 5, 17, 85
15) 1, 2, 3, 4, 6, 8, 12, 16, 24, 48
16) $2 \times 5 \times 5$
17) 5×5
18) 3×23
19) 3×7
20) $3 \times 3 \times 5$
21) $2 \times 2 \times 17$
22) 2×13
23) 2×43
24) 3×31

Greatest Common Factor

1) 10
2) 2
3) 5
4) 4
5) 1
6) 3
7) 3

8) 2
9) 2
10) 1
11) 2
12) 3
13) 4
14) 10

15) 1
16) 5
17) 2
18) 5
19) 12
20) 2
21) 10

Least Common Multiple

1) 28
2) 15
3) 80
4) 68
5) 24
6) 24
7) 18

8) 30
9) 152
10) 63
11) 551
12) 42
13) 150
14) 8

15) 150
16) 108
17) 72
18) 72
19) 780
20) 120
21) 90

Divisibility Rules

1) 16
<u>2</u> 3 <u>4</u> 5 6 7 <u>8</u> 9 10

2) 10
<u>2</u> 3 4 <u>5</u> 6 7 8 9 <u>10</u>

3) 15
2 <u>3</u> 4 <u>5</u> 6 7 8 9 10

4) 28
<u>2</u> 3 <u>4</u> 5 6 <u>7</u> 8 9 10

5) 36
<u>2</u> <u>3</u> <u>4</u> 5 <u>6</u> 7 8 <u>9</u> 10

6) 18
<u>2</u> <u>3</u> 4 5 <u>6</u> 7 8 <u>9</u> 10

7) 27
2 <u>3</u> 4 5 6 7 8 <u>9</u> 10

8) 70
<u>2</u> 3 4 <u>5</u> 6 <u>7</u> 8 9 <u>10</u>

9) 57
2 <u>3</u> 4 5 6 7 8 9 10

10) 102
<u>2</u> <u>3</u> 4 5 <u>6</u> 7 8 9 10

11) 144
<u>2</u> <u>3</u> <u>4</u> 5 <u>6</u> 7 <u>8</u> <u>9</u> 10

12) 75
2 <u>3</u> 4 <u>5</u> 6 7 8 9 10

Chapter 2: Integers and Absolute Value

Topics that you'll learn in this chapter:

✓ Adding and Subtracting Integers

✓ Multiplying and Dividing Integers

✓ Ordering Integers and Numbers

✓ Arrange, Order, and Comparing Integers

✓ Order of Operations

✓ Mixed Integer Computations

✓ Absolute Value

✓ Integers and Absolute Value

✓ Classifying Real Numbers Venn Diagram

Adding and Subtracting Integers

Helpful	-	**Integers:** {... , –3, –2, –1, 0, 1, 2, 3, ...} Includes: zero, counting numbers, and the negative of the counting numbers.	**Example:**
Hints		– Add a positive integer by moving to the right on the number line.	$12 + 10 = 22$ $25 - 13 = 12$
		– Add a negative integer by moving to the left on the number line.	$(-24) + 12 = -12$ $(-14) + (-12) = -26$
		– Subtract an integer by adding its opposite.	$14 - (-13) = 27$

✎ *Find the sum.*

1) $(- 12) + (- 4)$

2) $5 + (- 24)$

3) $(- 14) + 23$

4) $(- 8) + (39)$

5) $43 + (-12)$

6) $(- 23) + (- 4) + 3$

7) $4 + (- 12) + (- 10) + (- 25)$

8) $19 + (- 15) + 25 + 11$

9) $(- 9) + (- 12) + (32 - 14)$

10) $4 + (- 30) + (45 - 34)$

✎ *Find the difference.*

11) $(- 14) - (- 9) - (18)$

12) $(- 9) - (- 25)$

13) $(- 12) - (8)$

14) $(28) - (- 4)$

15) $(34) - (2)$

16) $(55) - (- 5) + (- 4)$

17) $(9) - (2) - (- 5)$

18) $(2) - (4) - (- 15)$

19) $(23) - (4) - (- 34)$

20) $(- 45) - (- 87)$

Multiplying and Dividing Integers

Helpful	(negative) × (negative) = positive	**Examples:**
	(negative) ÷ (negative) = positive	$3 \times 2 = 6$
Hints	(negative) × (positive) = negative	$3 \times -3 = -9$
	(negative) ÷ (positive) = negative	$-2 \times -2 = 4$
	(positive) × (positive) = positive	$10 \div 2 = 5$
		$-4 \div 2 = -2$
		$-12 \div -6 = 3$

✎ *Find each product.*

1) $(-8) \times (-2)$

2) 3×6

3) $(-4) \times 5 \times (-6)$

4) $2 \times (-6) \times (-6)$

5) $11 \times (-12)$

6) $10 \times (-5)$

7) 8×8

8) $(-8) \times (-9)$

9) $6 \times (-5) \times 3$

10) $6 \times (-1) \times 2$

✎ *Find each quotient.*

11) $18 \div 3$

12) $(-24) \div 4$

13) $(-63) \div (-9)$

14) $54 \div 9$

15) $20 \div (-2)$

16) $(-66) \div (-11)$

17) $64 \div 8$

18) $(-121) \div 11$

19) $72 \div 9$

20) $16 \div 4$

Ordering Integers and Numbers

Helpful *Hints*	To compare numbers, you can use number line! As you move from left to right on the number line, you find a bigger number!	**Example:** Order integers from least to greatest. $(-11, -13, 7, -2, 12)$ $-13 <-11< -2 < 7 <12$

✏️ **Order each set of integers from least to greatest.**

1) $-15, -19, 20, -4, 1$ ___, ___, ___, ___, ___, ___

2) $6, -5, 4, -3, 2$ ___, ___, ___, ___, ___, ___

3) $15, -42, 19, 0, -22$ ___, ___, ___, ___, ___, ___

4) $26, -91, 0, -13, 67, -55$ ___, ___, ___, ___, ___, ___

5) $-17, -71, 90, -25, -54, -39$ ___, ___, ___, ___, ___, ___

6) $98, 5, 46, 19, 77, 24$ ___, ___, ___, ___, ___, ___

✏️ **Order each set of integers from greatest to least.**

7) $-2, 5, -3, 6, -4$ ___, ___, ___, ___, ___, ___

8) $-37, 7, -17, 27, 47$ ___, ___, ___, ___, ___, ___

9) $32, -27, 19, -17, 15$ ___, ___, ___, ___, ___, ___

10) $68, 81, 21, -18, 94, 72$ ___, ___, ___, ___, ___, ___

Arrange, Order, and Comparing Integers

Helpful	When using a number line, numbers increase as you move to the right.	**Examples:**
Hints		$5 < 7,$ $-5 < -2$ $-18 < -12$

✏️*Arrange these integers in descending order.*

1) 21, 71, − 18, − 10, 82   ___, ___, ___, ___, ___, ___

2) 15, 11, 20, 12, − 9, − 5   ___, ___, ___, ___, ___, ___

3) − 5, 20, 15, 9, −11   ___, ___, ___, ___, ___, ___

4) 19, 18, − 9, − 6, − 11   ___, ___, ___, ___, ___, ___

5) 56, − 34, − 12, − 5, 32   ___, ___, ___, ___, ___, ___

✏️*Compare. Use >, =, <*

6) − 8 ____ 12   11) − 56 ____ − 58

7) − 10 ____ −16   12) 78 ____ 87

8) 43 ____ 34   13) − 92 ____ − 102

9) 15 ____ −16   14) − 12 ____ − 12

10) − 354 ____ −345   15) − 721 ____ − 821

Order of Operations

Helpful	-	Use "order of operations" rule when there are more than one math operation.	**Example:**
Hints	-	PEMDAS (parentheses / exponents / multiply / divide / add / subtract)	$(12 + 4) \div (-4) = -4$

✎ *Evaluate each expression.*

1) $(2 \times 2) + 5$

2) $24 - (3 \times 3)$

3) $(6 \times 4) + 8$

4) $25 - (4 \times 2)$

5) $(6 \times 5) + 3$

6) $64 - (2 \times 4)$

7) $25 + (1 \times 8)$

8) $(6 \times 7) + 7$

9) $48 \div (4 + 4)$

10) $(7 + 11) \div (-2)$

11) $9 + (2 \times 5) + 10$

12) $(5 + 8) \times \frac{3}{5} + 2$

13) $2 \times 7 - (\frac{10}{9 - 4})$

14) $(12 + 2 - 5) \times 7 - 1$

15) $(\frac{7}{5 - 1}) \times (2 + 6) \times 2$

16) $20 \div (4 - (10 - 8))$

17) $\frac{50}{4(5 - 4) - 3}$

18) $2 + (8 \times 2)$

Mixed Integer Computations

Helpful Hints	It worth remembering: (negative) × (negative) = positive (negative) ÷ (negative) = positive (negative) × (positive) = negative (negative) ÷ (positive) = negative (positive) × (positive) = positive	Example: $(-5) + 6 = 1$ $(-3) \times (-2) = 6$ $(9) \div (-3) = -3$

✎ Compute.

1) $(-70) \div (-5)$

2) $(-14) \times 3$

3) $(-4) \times (-15)$

4) $(-65) \div 5$

5) $18 \times (-7)$

6) $(-12) \times (-2)$

7) $\dfrac{(-60)}{(-20)}$

8) $24 \div (-8)$

9) $22 \div (-11)$

10) $\dfrac{(-27)}{3}$

11) $4 \times (-4)$

12) $\dfrac{(-48)}{12}$

13) $(-14) \times (-2)$

14) $(-7) \times (7)$

15) $\dfrac{-30}{-6}$

16) $(-54) \div 6$

17) $(-60) \div (-5)$

18) $(-7) \times (-12)$

19) $(-14) \times 5$

20) $88 \div (-8)$

Absolute Value

<table>
<tr>
<td>Helpful

Hints</td>
<td>Refers to the distance of a number from 0, the distances are positive. Therefore, absolute value of a number cannot be negative. $|-22| = 22$

$|x| = \begin{cases} x & for \ x \geq 0 \\ -x & for \ x < 0 \end{cases}$

$|x| < n \quad \Rightarrow -n < x < n$

$|x| > n \quad \Rightarrow x < -n \ or \ x > n$</td>
<td>Example:

$|12| \times |-2| = 24$</td>
</tr>
</table>

✎ Evaluate.

1) $|-4| + |-12| - 7$

2) $|-5| + |-13|$

3) $-18 + |-5 + 3| - 8$

4) $|27| \div |9|$

5) $|-9| \div |-1|$

6) $|200| \div |-100|$

7) $|55| \div |11|$

8) $|36| \div |-6|$

9) $|25| \times |-5|$

10) $|-3| \times |-8|$

11) $|12| \times |-5|$

12) $|11| \times |-6|$

13) $|-8| \times |4|$

14) $|-9| \times |-7|$

15) $|43 - 67 + 9| + |-11| - 1$

16) $|-45 + 78| + |23| - |45|$

17) $75 + |-11 - 30| - |2|$

18) $|-3 + 15| + |9 + 4| - 1$

Integers and Absolute Value

Helpful	To find an absolute value of a number, just find it's distance from 0!	**Example:**
Hints		$\|-6\| = 6$
		$\|6\| = 6$
		$\|-12\| = 12$
		$\|12\| = 12$

✎ *Write absolute value of each number.*

1) -4

2) -7

3) -8

4) 4

5) 5

6) -10

7) 1

8) 6

9) 8

10) -2

11) -1

12) 10

13) 3

14) 7

15) -5

16) -3

17) -9

18) 2

19) 4

20) -6

21) 9

✎ *Evaluate.*

22) $\|-43\| - \|12\| + 10$

23) $76 + \|-15 - 45\| - \|3\|$

24) $30 + \|-62\| - 46$

25) $\|32\| - \|-78\| + 90$

26) $\|-35 + 4\| + 6 - 4$

27) $\|-4\| + \|-11\|$

28) $\|-6 + 3 - 4\| + \|7 + 7\|$

29) $\|-9\| + \|-19\| - 5$

Classifying Real Numbers Venn Diagram

Helpful	**Natural numbers (counting numbers):** are the numbers that are used for counting. 1, 2, 3, ..., 100, ... are natural numbers.
Hints	**Whole numbers** are the natural numbers plus zero.
	Integers include all whole numbers plus "negatives" of the natural numbers.
Example: 0.25 =	**Rational numbers** are numbers that can be written as a fraction. Both top and bottom numbers must be integers.
rational number	**Irrational numbers** are all numbers which cannot be written as fractions.
and	**Real numbers** include both the rational and irrational numbers.
real number	

✍️*Identify all of the subsets of real number system to which each number belongs.*

Example:

0.1259 : Rational number

$\sqrt{2}$: Irrational number

3 : Natural number, whole number, Integer, rational number

1) 0

2) -5

3) -8.5

4) $\sqrt{4}$

5) -10

6) 18

7) 6

8) π

9) $1\frac{2}{7}$

10) -1

11) $\sqrt{5}$

Answers of Worksheets – Chapter 2

Adding and Subtracting Integers

1) -16

2) -19

3) 9

4) 31

5) 31

6) -24

7) -43

8) 40

9) -3

10) -15

11) -23

12) 16

13) -20

14) 32

15) 32

16) 56

17) 12

18) 13

19) 53

20) 42

Multiplying and Dividing Integers

1) 16

2) 18

3) 120

4) 72

5) -132

6) -50

7) 64

8) 72

9) -90

10) -12

11) 6

12) -6

13) 7

14) 6

15) -10

16) 6

17) 8

18) -11

19) 8

20) 4

Ordering Integers and Numbers

1) $-19, -15, -4, 1, 20$

2) $-5, -3, 2, 4, 6$

3) $-42, -22, 0, 15, 19$

4) $-91, -55, -13, 0, 26, 67$

5) $-71, -54, -39, -25, -17, 90$

6) $5, 19, 24, 46, 77, 98$

7) $6, 5, -2, -3, -4$

8) $47, 27, 7, -17, -37$

9) $32, 19, 15, -17, -27$

10) $94, 81, 72, 68, 21, -18$

Arrange and Order, Comparing Integers

1) 82, 71, 21, − 10, − 18

2) 20, 15, 12, 11, − 5, − 9

3) 20, 15, 9, − 5, −11

4) 19, 18, − 6, − 9, − 11

5) 56, 32, − 5, − 12, − 34

6) <	10) <	14) =
7) >	11) >	15) >
8) >	12) <	
9) >	13) >	

Order of Operations

1) 9	7) 33	13) 12
2) 15	8) 49	14) 62
3) 32	9) 6	15) 28
4) 17	10) − 9	16) 10
5) 33	11) 29	17) 50
6) 56	12) 9.8	18) 18

Mixed Integer Computations

1) 14	8) − 3	15) 5
2) − 42	9) − 2	16) − 9
3) 60	10) − 9	17) 12
4) − 13	11) − 16	18) 84
5) − 126	12) − 4	19) − 70
6) 24	13) 28	20) − 11
7) 3	14) − 49	

Absolute Value

1) 9	7) 5	13) 32
2) 18	8) 6	14) 63
3) − 24	9) 125	15) 25
4) 3	10) 24	16) 11
5) 9	11) 60	17) 114
6) 2	12) 66	18) 24

Integers and Absolute Value

1) 4	11) 1	21) 9
2) 7	12) 10	22) 41
3) 8	13) 3	23) 133
4) 4	14) 7	24) 46
5) 5	15) 5	25) 44
6) 10	16) 3	26) 33
7) 1	17) 9	27) 15
8) 6	18) 2	28) 21
9) 8	19) 4	29) 23
10) 2	20) 6	

Classifying Real Numbers Venn Diagram

1) 0: whole number, integer, rational number
2) − 5: integer, rational number
3) − 8.5: rational number
4) $\sqrt{4}$: natural number, whole number, integer, rational number
5) − 10: integer, rational number
6) 18 : natural number, whole number, integer, rational number
7) 6: natural number, whole number, integer, rational number
8) π: irrational number
9) $1\frac{2}{7}$: rational number
10)   − 1: integer, rational number
11) $\sqrt{5}$: irrational number

Chapter 3: Percent

Topics that you'll learn in this chapter:

- ✓ Percentage Calculations
- ✓ Converting Between Percent, Fractions, and Decimals
- ✓ Percent Problems
- ✓ Find What Percentage a Number Is of Another
- ✓ Find a Percentage of a Given Number
- ✓ Percent of Increase and Decrease

Percentage Calculations

Helpful	-	Use the following formula to find part, whole, or percent: $$part = \frac{percent}{100} \times whole$$	**Example:** $\frac{20}{100} \times 100 = 20$
Hints			

✎ *Calculate the percentages.*

1) 50% of 25

2) 80% of 15

3) 30% of 34

4) 70% of 45

5) 10% of 0

6) 80% of 22

7) 65% of 8

8) 78% of 54

9) 50% of 80

10) 20% of 10

11) 40% of 40

12) 90% of 0

13) 20% of 70

14) 55% of 60

15) 80% of 10

16) 20% of 880

17) 70% of 100

18) 80% of 90

✎ *Solve.*

19) 50 is what percentage of 75?

20) What percentage of 100 is 70

21) Find what percentage of 60 is 35.

22) 40 is what percentage of 80?

Converting Between Percent, Fractions, and Decimals

Helpful	– To a percent: Move the decimal point 2 places to the right and add the % symbol.	**Examples:**
Hints	– Divide by 100 to convert a number from percent to decimal.	30% = 0.3
		0.24 = 24%

✍ *Converting fractions to decimals.*

1) $\dfrac{50}{100}$

2) $\dfrac{38}{100}$

3) $\dfrac{15}{100}$

4) $\dfrac{80}{100}$

5) $\dfrac{7}{100}$

6) $\dfrac{35}{100}$

7) $\dfrac{90}{100}$

8) $\dfrac{20}{100}$

9) $\dfrac{7}{100}$

✍ *Write each decimal as a percent.*

10) 0.5

11) 0.9

12) 0.002

13) 0.524

14) 0.1

15) 0.03

16) 3.63

17) 0.008

18) 4.78

Percent Problems

Helpful	Base = Part ÷ Percent	**Example:**
	Part = Percent × Base	2 is 10% of 20.
Hints	Percent = Part ÷ Base	2 ÷ 0.10 = 20
		2 = 0.10 × 20
		0.10 = 2 ÷ 20

✐*Solve each problem.*

1) 51 is 340% of what?

2) 93% of what number is 97?

3) 27% of 142 is what number?

4) What percent of 125 is 29.3?

5) 60 is what percent of 126?

6) 67 is 67% of what?

7) 67 is 13% of what?

8) 41% of 78 is what?

9) 1 is what percent of 52.6?

10) What is 59% of 14 m?

11) What is 90% of 130 inches?

12) 16 inches is 35% of what?

13) 90% of 54.4 hours is what?

14) What percent of 33.5 is 21?

15) Liam scored 22 out of 30 marks in Algebra, 35 out of 40 marks in science and 89 out of 100 marks in mathematics. In which subject his percentage of marks in best?

16) Ella require 50% to pass. If she gets 280 marks and falls short by 20 marks, what were the maximum marks she could have got?

Find What Percentage a Number Is of Another

Helpful *Hints*	PERCENT: the number with the percent sign (%). PART: the number with the word "is". WHOLE: the number with the word "of". – Divide the Part by the Base. – Convert the answer to percent.	**Example:** 20 is what percent of 50? $20 \div 50 = 0.40 = 40\%$

Find the percentage of the numbers.

1) 5 is what percent of 90?

2) 15 is what percent of 75?

3) 20 is what percent of 400?

4) 18 is what percent of 90?

5) 3 is what percent of 15?

6) 8 is what percent of 80?

7) 11 is what percent of 55?

8) 9 is what percent of 90?

9) 2.5 is what percent of 10?

10) 5 is what percent of 25?

11) 60 is what percent of 20?

12) 12 is what percent of 48?

13) 14 is what percent of 28?

14) 8.2 is what percent of 32.8?

15) 1200 is what percent of 4,800?

16) 4,000 is what percent of 20,000?

17) 45 is what percent of 900?

18) 10 is what percent of 200?

19) 15 is what percent of 60?

20) 1.2 is what percent of 24?

Find a Percentage of a Given Number

Helpful	-	Use following formula to find part, whole, or percent: $\text{part} = \dfrac{\text{percent}}{100} \times \text{whole}$	Example: $\dfrac{50}{100} \times 50 = 25$
Hints			

Find a Percentage of a Given Number.

1) 90% of 50

2) 40% of 50

3) 10% of 0

4) 80% of 80

5) 60% of 40

6) 50% of 60

7) 30% of 20

8) 35% of 10

9) 10% of 80

10) 10% of 60

11) 100% 0f 50

12) 90% of 34

13) 80% of 42

14) 90% of 12

15) 20% of 56

16) 40% of 40

17) 40% of 6

18) 70% of 38

19) 30% of 3

20) 40% of 50

21) 100% of 8

Percent of Increase and Decrease

Helpful *Hints*	– To find the percentage increase:	**Example:**
	New Number – Original Number	From 84 miles to 24 miles =
	The result ÷ Original Number × 100	71.43% decrease
	If your answer is a negative number, then this is a percentage decrease.	
	To calculate percentage decrease:	
	Original Number – New Number	
	The result ÷ Original Number × 100	

Find each percent change to the nearest percent. Increase or decrease.

1) From 32 grams to 82 grams.

2) From 150 m to 45 m

3) From $438 to $443

4) From 256 ft to 140 ft

5) From 6469 ft to 7488 ft

6) From 36 inches to 90 inches

7) From 54 ft to 104 ft

8) From 84 miles to 24 miles

9) The population of a place in a particular year increased by 15%. Next year it decreased by 15%. Find the net increase or decrease percent in the initial population.

10) The salary of a doctor is increased by 40%. By what percent should the new salary be reduced in order to restore the original salary?

Answers of Worksheets – Chapter 3

Percentage Calculations

1) 12.5
2) 12
3) 10.2
4) 31.5
5) 0
6) 17.6
7) 5.2
8) 42.12

9) 40
10) 2
11) 16
12) 0
13) 14
14) 33
15) 8
16) 176

17) 70
18) 72
19) 67%
20) 70%
21) 58%
22) 50%

Converting Between Percent, Fractions, and Decimals

1) 0.5
2) 0.38
3) 0.15
4) 0.8
5) 0.07
6) 0.35

7) 0.9
8) 0.2
9) 0.07
10) 50%
11) 90%
12) 0.2%

13) 52.4%
14) 10%
15) 3%
16) 363%
17) 0.8%
18) 478%

Percent Problems

1) 15
2) 104.3
3) 38.34
4) 23.44%
5) 47.6%
6) 100

7) 515.4
8) 31.98
9) 1.9%
10) 8.3 m
11) 117 inches
12) 45.7inches

13) 49 hours
14) 62.7%
15) Mathematics
16) 600

Find What Percentage a Number Is of Another

1) 45 is what percent of 90? 50 %

2) 15 is what percent of 75? 20 %

3) 20 is what percent of 400? 5 %

4) 18 is what percent of 90? 20 %

5) 3 is what percent of 15? 20 %

6) 8 is what percent of 80? 10 %

7) 11 is what percent of 55? 20 %

8) 9 is what percent of 90? 10 %

9) 2.5 is what percent of 10? 25 %

10) 5 is what percent of 25? 20 %

11) 60 is what percent of 20? 300 %

12) 12 is what percent of 48? 25 %

13) 14 is what percent of 28? 50 %

14) 8.2 is what percent of 32.8? 25 %

15) 1200 is what percent of 4,800?
    25 %

16) 4,000 is what percent of 20,000?
    20 %

17) 45 is what percent of 900? 5 %

18) 10 is what percent of 200? 5 %

19) 15 is what percent of 60? 25 %

20) 1.2 is what percent of 24? 5 %

Find a Percentage of a Given Number

1) 45

2) 20

3) 0

4) 64

5) 24

6) 30

7) 6

8) 3.5

9) 8

10) 6

11) 50

12) 30.6

13) 33.6

14) 10.8

15) 11.2

16) 16

17) 2.4

18) 26.6

19) 0.9

20) 20

21) 8

Percent of Increase and Decrease

1) 156.25% increase

2) 70% decrease

3) 1.142% increase

4) 45.31% decrease

5) 15.75% increase

6) 150% increase

7) 92.6% increase

8) 71.43% decrease

9) 2.25% decrease

10) $28\frac{4}{7}\%$

Chapter 4: Algebraic Expressions

Topics that you'll learn in this chapter:

- ✓ Expressions and Variables
- ✓ Simplifying Variable Expressions
- ✓ Translate Phrases into an Algebraic Statement
- ✓ The Distributive Property
- ✓ Evaluating One Variable
- ✓ Evaluating Two Variables
- ✓ Combining like Terms

Expressions and Variables

Helpful *Hints*	A variable is a letter that represents unknown numbers. A variable can be used in the same manner as all other numbers:

Addition	$2 + a$	2 plus a
Subtraction	$y - 3$	y minus 3
Division	$\dfrac{4}{x}$	4 divided by x
Multiplication	5a	5 times a

✍ *Simplify each expression.*

1) $x + 5x$,

    use $x = 5$

2) $8\,(-3x + 9) + 6$,

    use $x = 6$

3) $10x - 2x + 6 - 5$,

    use $x = 5$

4) $2x - 3x - 9$,

    use $x = 7$

5) $(-6)\,(-2x - 4y)$,

    use $x = 1$, $y = 3$

6) $8x + 2 + 4\,y$,

    use $x = 9$, $y = 2$

7) $(-6)\,(-8x - 9y)$,

    use $x = 5$, $y = 5$

8) $6x + 5y$,

    use $x = 7$, $y = 4$

✍ *Simplify each expression.*

9) $5\,(-4 + 2x)$

10) $-3 - 5x - 6x + 9$

11) $6x - 3x - 8 + 10$

12) $(-8)\,(6x - 4) + 12$

13) $9\,(7x + 4) + 6x$

14) $(-9)\,(-5x + 2)$

Simplifying Variable Expressions

Helpful	– Combine "like" terms. (values with same variable and same power)	**Example:**
Hints	– Use distributive property if necessary.	$2x + 2 (1 - 5x) =$
	Distributive Property:	$2x + 2 - 10x = -8x + 2$
	$a (b + c) = ab + ac$	

✍ *Simplify each expression.*

1) $-2 - x^2 - 6x^2$

2) $3 + 10x^2 + 2$

3) $8x^2 + 6x + 7x^2$

4) $5x^2 - 12x^2 + 8x$

5) $2x^2 - 2x - x$

6) $(-6)(8x - 4)$

7) $4x + 6(2 - 5x)$

8) $10x + 8(10x - 6)$

9) $9(-2x - 6) - 5$

10) $3(x + 9)$

11) $7x + 3 - 3x$

12) $2.5x^2 \times (-8x)$

✍ *Simplify.*

13) $-2(4 - 6x) - 3x$, $x = 1$

14) $2x + 8x$, $x = 2$

15) $9 - 2x + 5x + 2$, $x = 5$

16) $5(3x + 7)$, $x = 3$

17) $2(3 - 2x) - 4$, $x = 6$

18) $5x + 3x - 8$, $x = 3$

19) $x - 7x$, $x = 8$

20) $5(-2 - 9x)$, $x = 4$

Translate Phrases into an Algebraic Statement

Helpful	**Translating key words and phrases into algebraic expressions:**
	Addition: plus, more than, the sum of, etc.
Hints	**Subtraction:** minus, less than, decreased, etc.
	Multiplication: times, product, multiplied, etc.
	Division: quotient, divided, ratio, etc.
	Example:
	eight more than a number is 20
	$8 + x = 20$

✎ *Write an algebraic expression for each phrase.*

1) A number increased by forty–two.

2) The sum of fifteen and a number

3) The difference between fifty–six and a number.

4) The quotient of thirty and a number.

5) Twice a number decreased by 25.

6) Four times the sum of a number and − 12.

7) A number divided by − 20.

8) The quotient of 60 and the product of a number and − 5.

9) Ten subtracted from a number.

10) The difference of six and a number.

The Distributive Property

Helpful	Distributive Property:	Example:
Hints	$a(b+c) = ab + ac$	$3(4+3x)$ $= 12 + 9x$

✎ *Use the distributive property to simply each expression.*

1) $-(-2 - 5x)$

2) $(-6x + 2)(-1)$

3) $(-5)(x - 2)$

4) $-(7 - 3x)$

5) $8(8 + 2x)$

6) $2(12 + 2x)$

7) $(-6x + 8)4$

8) $(3 - 6x)(-7)$

9) $(-12)(2x + 1)$

10) $(8 - 2x)9$

11) $(-2x)(-1 + 9x) - 4x(4 + 5x)$

12) $3(-5x - 3) + 4(6 - 3x)$

13) $(-2)(x + 4) - (2 + 3x)$

14) $(-4)(3x - 2) + 6(x + 1)$

15) $(-5)(4x - 1) + 4(x + 2)$

16) $(-3)(x + 4) - (2 + 3x)$

Evaluating One Variable

Helpful	− To evaluate one variable expression, find the variable and substitute a number for that variable.	**Example:**
Hints	− Perform the arithmetic operations.	$4x + 8, x = 6$ $4(6) + 8 = 24 + 8 = 32$

✍ *Simplify each algebraic expression.*

1) $9 - x$, $x = 3$

2) $x + 2$, $x = 5$

3) $3x + 7$, $x = 6$

4) $x + (-5)$, $x = -2$

5) $3x + 6$, $x = 4$

6) $4x + 6$, $x = -1$

7) $10 + 2x - 6$, $x = 3$

8) $10 - 3x$, $x = 8$

9) $\dfrac{20}{x} - 3$, $x = 5$

10) $(-3) + \dfrac{x}{4} + 2x$, $x = 16$

11) $(-2) + \dfrac{x}{7}$, $x = 21$

12) $(-\dfrac{14}{x}) - 9 + 4x$, $x = 2$

13) $(-\dfrac{6}{x}) - 9 + 2x$, $x = 3$

14) $(-2) + \dfrac{x}{8}$, $x = 16$

Evaluating Two Variables

Helpful *Hints*	To evaluate an algebraic expression, substitute a number for each variable and perform the arithmetic operations.	**Example:** $2x + 4y - 3 + 2,$ $x = 5, y = 3$ $2(5) + 4(3) - 3 + 2$ $= 10$ $+ 12 - 3 + 2$ $= 21$

✍ *Simplify each algebraic expression.*

1) $2x + 4y - 3 + 2,$

   $x = 5, y = 3$

2) $(-\dfrac{12}{x}) + 1 + 5y,$

   $x = 6, y = 8$

3) $(-4)(-2a - 2b),$

   $a = 5, b = 3$

4) $10 + 3x + 7 - 2y,$

   $x = 7, y = 6$

5) $9x + 2 - 4y,$

   $x = 7, y = 5$

6) $6 + 3(-2x - 3y),$

   $x = 9, y = 7$

7) $12x + y,$

   $x = 4, y = 8$

8) $x \times 4 \div y,$

   $x = 3, y = 2$

9) $2x + 14 + 4y,$

   $x = 6, y = 8$

10) $4a - (5 - b),$

   $a = 4, b = 6$

Combining like Terms

Helpful *Hints*	− Terms are separated by "+" and "−" signs. − Like terms are terms with same variables and same powers. − Be sure to use the "+" or "−" that is in front of the coefficient.	**Example:** $22x + 6 + 2x =$ $24x + 6$

✏️ *Simplify each expression.*

1) $5 + 2x − 8$

2) $(− 2x + 6)\ 2$

3) $7 + 3x + 6x − 4$

4) $(− 4) − (3)(5x + 8)$

5) $9x − 7x − 5$

6) $x − 12x$

7) $7\ (3x + 6) + 2x$

8) $(− 11x) − 10x$

9) $3x − 12 − 5x$

10) $13 + 4x − 5$

11) $(− 22x) + 8x$

12) $2\ (4 + 3x) − 7x$

13) $(− 4x) − (6 − 14x)$

14) $5\ (6x − 1) + 12x$

15) $22x + 6 + 2x$

16) $(− 13x) − 14x$

17) $(− 6x) − 9 + 15x$

18) $(− 6x) + 7x$

19) $(− 5x) + 12 + 7x$

20) $(− 3x) − 9 + 15x$

21) $20x − 19x$

Answers of Worksheets – Chapter 4

Expressions and Variables

1) 30
2) −66
3) 41
4) −16
5) 84

6) 82
7) 510
8) 62
9) 10x − 20
10) 6 − 11x

11) 3x + 2
12) 44 − 48x
13) 69x + 36
14) 45x − 18

Simplifying Variable Expressions

1) $-7x^2 - 2$
2) $10x^2 + 5$
3) $15x^2 + 6x$
4) $-7x^2 + 8x$
5) $2x^2 - 3x$
6) $-48x + 24$
7) $-26x + 12$

8) 90x − 48
9) −18x − 59
10) 3x + 27
11) 4x + 3
12) $-20x^3$
13) 1
14) 20

15) 26
16) 80
17) − 22
18) 16
19) − 48
20) − 190

Translate Phrases into an Algebraic Statement

1) $x + 42$
2) $15 + x$
3) $56 - x$
4) $30/x$
5) 2x − 25
6) $4(x + (-12))$
7) $\dfrac{x}{-20}$
8) $\dfrac{60}{-5x}$
9) $x - 10$
10) $6 - x$

The Distributive Property

1) 5x + 2
2) 6x − 2
3) −5x + 10
4) 3x − 7
5) 16x + 64
6) 4x + 24

7) − 24x + 32
8) 42x − 21
9) − 24x − 12
10) − 18x + 72
11) $-38x^2 - 14x$
12) − 27x + 15

13) − 5x − 10
14) − 6x + 14
15) − 16x + 13
16) − 6x − 14

Evaluating One Variable

1) 6
2) 7
3) 25
4) −7
5) 18

6) 2
7) 10
8) −14
9) 1
10) 33

11) 1
12) −8
13) −5
14) 0
15) −176

Evaluating Two Variables

1) 21
2) 39
3) 64
4) 26

5) 45
6) −111
7) 56
8) 6

9) 58
10) 17

Combining like Terms

1) 2x − 3
2) −4x + 12
3) 9x + 3
4) −15x − 28
5) 2x − 5
6) −11x
7) 23x + 42

8) −21x
9) −2x − 12
10) 4x + 8
11) −14x
12) − x + 8
13) 10x − 6
14) 42x − 5

15) 24x + 6
16) −27x
17) 9x − 9
18) x
19) 2x + 12
20) 12x − 9
21) x

Chapter 5: Equations

Topics that you'll learn in this chapter:

- ✓ One–Step Equations
- ✓ One–Step Equation Word Problems
- ✓ Two–Step Equations
- ✓ Two–Step Equation Word Problems
- ✓ Multi–Step Equations

One–Step Equations

Helpful	-	The values of two expressions on both sides of an equation are equal.	**Example:**
Hints		$ax + b = c$	$-8x = 16$
	-	You only need to perform one Math operation in order to solve the equation.	$x = -2$

✎ *Solve each equation.*

1) x + 3 = 17

2) 22 = (− 8) + x

3) 3x = (− 30)

4) (− 36) = (− 6x)

5) (− 6) = 4 + x

6) 2 + x = (− 2)

7) 20x = (− 220)

8) 18 = x + 5

9) (− 23) + x = (− 19)

10) 5x = (− 45)

11) x − 12 = (− 25)

12) x − 3 = (− 12)

13) (− 35) = x − 27

14) 8 = 2x

15) (− 6x) = 36

16) (− 55) = (− 5x)

17) x − 30 = 20

18) 8x = 32

19) 36 = (− 4x)

20) 4x = 68

21) 30x = 300

One–Step Equation Word Problems

Helpful	– Define the variable.
	– Translate key words and phrases into math equation.
Hints	– Isolate the variable and solve the equation.

✎*Solve.*

1) How many boxes of envelopes can you buy with $18 if one box costs $3?

2) After paying $6.25 for a salad, Ella has $45.56. How much money did she have before buying the salad?

3) How many packages of diapers can you buy with $50 if one package costs $5?

4) Last week James ran 20 miles more than Michael. James ran 56 miles. How many miles did Michael run?

5) Last Friday Jacob had $32.52. Over the weekend he received some money for cleaning the attic. He now has $44. How much money did he receive?

6) After paying $10.12 for a sandwich, Amelia has $35.50. How much money did she have before buying the sandwich?

Two–Step Equations

Helpful *Hints*	– You only need to perform two math operations (add, subtract, multiply, or divide) to solve the equation. – Simplify using the inverse of addition or subtraction. – Simplify further by using the inverse of multiplication or division.	**Example:** $-2(x-1) = 42$ $(x-1) = -21$ $x = -20$

✎ *Solve each equation.*

1) $5(8+x) = 20$

2) $(-7)(x-9) = 42$

3) $(-12)(2x-3) = (-12)$

4) $6(1+x) = 12$

5) $12(2x+4) = 60$

6) $7(3x+2) = 42$

7) $8(14+2x) = (-34)$

8) $(-15)(2x-4) = 48$

9) $3(x+5) = 12$

10) $\dfrac{3x-12}{6} = 4$

11) $(-12) = \dfrac{x+15}{6}$

12) $110 = (-5)(2x-6)$

13) $\dfrac{x}{8} - 12 = 4$

14) $20 = 12 + \dfrac{x}{4}$

15) $\dfrac{-24+x}{6} = (-12)$

16) $(-4)(5+2x) = (-100)$

17) $(-12x) + 20 = 32$

18) $\dfrac{-2+6x}{4} = (-8)$

19) $\dfrac{x+6}{5} = (-5)$

20) $(-9) + \dfrac{x}{4} = (-15)$

Two–Step Equation Word Problems

Helpful	– Translate the word problem into equations with variables.
Hints	– Solve the equations to find the solutions to the word problems.

✍️ *Solve.*

1) The sum of three consecutive even numbers is 48. What is the smallest of these numbers?

2) How old am I if 400 reduced by 2 times my age is 244?

3) For a field trip, 4 students rode in cars and the rest filled nine buses. How many students were in each bus if 472 students were on the trip?

4) The sum of three consecutive numbers is 72. What is the smallest of these numbers?

5) 331 students went on a field trip. Six buses were filled, and 7 students traveled in cars. How many students were in each bus?

6) You bought a magazine for $5 and four erasers. You spent a total of $25. How much did each eraser cost?

Multi–Step Equations

Helpful *Hints*	– Combine "like" terms on one side. – Bring variables to one side by adding or subtracting. – Simplify using the inverse of addition or subtraction. – Simplify further by using the inverse of multiplication or division.	**Example:** $3x + 15 = -2x + 5$ Add 2x both sides $5x + 15 = +5$ Subtract 15 both sides $5x = -10$ Divide by 5 both sides $x = -2$

✍ *Solve each equation.*

1) $-(2 - 2x) = 10$

2) $-12 = -(2x + 8)$

3) $3x + 15 = (-2x) + 5$

4) $-28 = (-2x) - 12x$

5) $2(1 + 2x) + 2x = -118$

6) $3x - 18 = 22 + x - 3 + x$

7) $12 - 2x = (-32) - x + x$

8) $7 - 3x - 3x = 3 - 3x$

9) $6 + 10x + 3x = (-30) + 4x$

10) $(-3x) - 8(-1 + 5x) = 352$

11) $24 = (-4x) - 8 + 8$

12) $9 = 2x - 7 + 6x$

13) $6(1 + 6x) = 294$

14) $-10 = (-4x) - 6x$

15) $4x - 2 = (-7) + 5x$

16) $5x - 14 = 8x + 4$

17) $40 = -(4x - 8)$

18) $(-18) - 6x = 6(1 + 3x)$

19) $x - 5 = -2(6 + 3x)$

20) $6 = 1 - 2x + 5$

Answers of Worksheets – Chapter 5

One–Step Equations

1) 14
2) 30
3) – 10
4) 6
5) – 10
6) – 4
7) – 11

8) 13
9) 4
10) – 9
11) – 13
12) – 9
13) – 8
14) 4

15) – 6
16) 11
17) 50
18) 4
19) – 9
20) 17
21) 10

One–Step Equation Word Problems

1) 6
2) $51.81

3) 10
4) 36

5) 11.48
6) 45.62

Two–Step Equations

1) – 4
2) 3
3) 2
4) 1
5) 0.5
6) $\frac{4}{3}$
7) $-\frac{73}{8}$

8) $\frac{2}{5}$
9) – 1
10) 12
11) – 87
12) – 8
13) 128
14) 32

15) – 48
16) 10
17) – 1
18) – 5
19) – 31
20) – 24

Two–Step Equation Word Problems

1) 14
2) 78

3) 52
4) 23

5) 54
6) $4

Multi–Step Equations

1) 6
2) 2
3) – 2
4) 2
5) – 20
6) 37
7) 22

8) $\frac{4}{3}$
9) – 4
10) – 8
11) – 6
12) 2
13) 8

14) 1
15) 5
16) – 6
17) – 8
18) – 1
19) – 1
20) 0

Chapter 6: Proportions and Ratios

Topics that you'll learn in this chapter:

- ✓ Writing Ratios
- ✓ Simplifying Ratios
- ✓ Proportional Ratios
- ✓ Create a Proportion
- ✓ Similar Figures
- ✓ Similar Figure Word Problems
- ✓ Ratio and Rates Word Problems

Writing Ratios

Helpful	− A ratio is a comparison of two numbers. Ratio can be written as a division.	**Example:**
Hints		$3:5$, or $\frac{3}{5}$

✍️ *Express each ratio as a rate and unite rate.*

1) 120 miles on 4 gallons of gas.

2) 24 dollars for 6 books.

3) 200 miles on 14 gallons of gas

4) 24 inches of snow in 8 hours

✍️ *Express each ratio as a fraction in the simplest form.*

5) 3 feet out of 30 feet

6) 18 cakes out of 42 cakes

7) 16 dimes t0 24 dimes

8) 12 dimes out of 48 coins

9) 14 cups to 84 cups

10) 45 gallons to 65 gallons

11) 10 miles out of 40 miles

12) 22 blue cars out of 55 cars

13) 32 pennies to 300 pennies

14) 24 beetles out of 86 insects

Proportional Ratios

$Helpful$ $Hints$	– A proportion means that two ratios are equal. It can be written in two ways: $\frac{a}{b} = \frac{c}{d}$, a : b = c : d	**Example:** $\frac{9}{3} = \frac{6}{d}$ $d = 2$

Solve each proportion.

1) $\frac{3}{6} = \frac{8}{d}$

2) $\frac{k}{5} = \frac{12}{15}$

3) $\frac{30}{5} = \frac{12}{x}$

4) $\frac{x}{2} = \frac{1}{8}$

5) $\frac{d}{3} = \frac{2}{6}$

6) $\frac{27}{7} = \frac{30}{x}$

7) $\frac{8}{5} = \frac{k}{15}$

8) $\frac{60}{20} = \frac{3}{d}$

9) $\frac{x}{3} = \frac{12}{18}$

10) $\frac{25}{5} = \frac{x}{8}$

11) $\frac{12}{x} = \frac{4}{2}$

12) $\frac{x}{4} = \frac{18}{2}$

13) $\frac{80}{10} = \frac{k}{10}$

14) $\frac{12}{6} = \frac{6}{d}$

15) $\frac{x}{4} = \frac{30}{5}$

16) $\frac{9}{5} = \frac{k}{5}$

17) $\frac{45}{15} = \frac{15}{d}$

18) $\frac{60}{x} = \frac{10}{3}$

19) $\frac{d}{3} = \frac{14}{6}$

20) $\frac{k}{4} = \frac{4}{2}$

21) $\frac{4}{2} = \frac{x}{7}$

Simplifying Ratios

Helpful *Hints*	– You can calculate equivalent ratios by multiplying or dividing both sides of the ratio by the same number.	**Examples:** 3 : 6 = 1 : 2 4 : 9 = 8 : 18

✍ *Reduce each ratio.*

1) 21 : 49	9) 35 : 45	17) 3 : 36
2) 20 : 40	10) 8 : 20	18) 8 : 16
3) 10 : 50	11) 25 : 35	19) 6 : 100
4) 14 : 18	12) 21 : 27	20) 2 : 20
5) 45 : 27	13) 52 : 82	21) 10 : 60
6) 49 : 21	14) 12 : 36	22) 14 : 63
7) 100 : 10	15) 24 : 3	23) 68 : 80
8) 12 : 8	16) 15 : 30	24) 8 : 80

Create a Proportion

Helpful	– A proportion contains 2 equal fractions! A proportion simply means that two fractions are equal.	**Example:**
Hints		2, 4, 8, 16
		$\dfrac{2}{4} = \dfrac{8}{16}$

✍ *Create proportion from the given set of numbers.*

1) 1, 6, 2, 3

2) 12, 144, 1, 12

3) 16, 4, 8, 2

4) 9, 5, 27, 15

5) 7, 10, 60, 42

6) 8, 7, 24, 21

7) 10, 5, 8, 4

8) 3, 12, 8, 2

9) 2, 2, 1, 4

10) 3, 6, 7, 14

11) 2, 6, 5, 15

12) 7, 2, 14, 4

Similar Figures

Helpful	– Two or more figures are similar if the corresponding angles are equal, and the corresponding sides are in proportion.	**Example:**
Hints		3–4–5 triangle is similar to a 6–8–10 triangle

✎ *Each pair of figures is similar. Find the missing side.*

1)

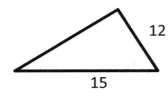

2)

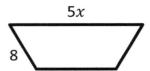

3)

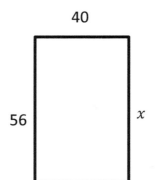

Similar Figure Word Problems

Helpful *Hints*	To solve a similarity word problem, create a proportion and use cross multiplication method!	**Example:** $\dfrac{x}{4} = \dfrac{8}{16}$ $16x = 4 \times 8$ $x = 2$

✎ *Answer each question and round your answer to the nearest whole number.*

1) If a 42.9 ft tall flagpole casts a 253.1 ft long shadow, then how long is the shadow that a 6.2 ft tall woman casts?

2) A model igloo has a scale of 1 in : 2 ft. If the real igloo is 10 ft wide, then how wide is the model igloo?

3) If a 18 ft tall tree casts a 9 ft long shadow, then how tall is an adult giraffe that casts a 7 ft shadow?

4) Find the distance between San Joe and Mount Pleasant if they are 2 cm apart on a map with a scale of 1 cm : 9 km.

5) A telephone booth that is 8 ft tall casts a shadow that is 4 ft long. Find the height of a lawn ornament that casts a 2 ft shadow.

Ratio and Rates Word Problems

Helpful	To solve a ratio or a rate word problem, create a proportion and use cross multiplication method!	**Example:**
Hints		$\frac{x}{4} = \frac{8}{16}$ $16x = 4 \times 8$ $x = 2$

✍ *Solve.*

1) In a party, 10 soft drinks are required for every 12 guests. If there are 252 guests, how many soft drink is required?

2) In Jack's class, 18 of the students are tall and 10 are short. In Michael's class 54 students are tall and 30 students are short. Which class has a higher ratio of tall to short students?

3) Are these ratios equivalent?
   12 cards to 72 animals 11 marbles to 66 marbles

4) The price of 3 apples at the Quick Market is $1.44. The price of 5 of the same apples at Walmart is $2.50. Which place is the better buy?

5) The bakers at a Bakery can make 160 bagels in 4 hours. How many bagels can they bake in 16 hours? What is that rate per hour?

6) You can buy 5 cans of green beans at a supermarket for $3.40. How much does it cost to buy 35 cans of green beans?

Answers of Worksheets – Chapter 6

Writing Ratios

1) $\frac{120\ miles}{4\ gallons}$, 30 miles per gallon

2) $\frac{24\ dollars}{6\ books}$, 4.00 dollars per book

3) $\frac{200\ miles}{14\ gallons}$, 14.29 miles per gallon

4) $\frac{24"\ of\ snow}{8\ hours}$, 3 inches of snow per hour

5) $\frac{1}{10}$

6) $\frac{3}{7}$

7) $\frac{2}{3}$

8) $\frac{1}{4}$

9) $\frac{1}{6}$

10) $\frac{9}{13}$

11) $\frac{1}{4}$

12) $\frac{2}{5}$

13) $\frac{8}{75}$

14) $\frac{12}{43}$

Simplifying Ratios

1) 3 : 7
2) 1 : 2
3) 1 : 5
4) 7 : 9
5) 5 : 3
6) 7 : 3
7) 10 : 1
8) 3 : 2

9) 7 : 9
10) 2 : 5
11) 5 : 7
12) 7 : 9
13) 26 : 41
14) 1 : 3
15) 8 : 1
16) 1 : 2

17) 1 : 12
18) 1 : 2
19) 3 : 50
20) 1 : 10
21) 1: 6
22) 2 : 9
23) 17 : 20
24) 1 : 10

Proportional Ratios

1) 16
2) 4
3) 2
4) 0.25

5) 1
6) 7.78
7) 24
8) 1

9) 2
10) 40
11) 6
12) 36

13) 80	16) 9	19) 7
14) 3	17) 5	20) 8
15) 24	18) 18	21) 14

Create a Proportion

1) 1 : 3 = 2 : 6	5) 7 : 42, 10 : 60	9) 4 : 2 = 2 : 1
2) 12 : 144 = 1 : 12	6) 7 : 21 = 8 : 24	10) 7 : 3 = 14 : 6
3) 2 : 4 = 8 : 16	7) 8 : 10 = 4 : 5	11) 5 : 2 = 15 : 6
4) 5 : 15 = 9 : 27	8) 2 : 3 = 8 : 12	12) 7 : 2 = 14 : 4

Similar Figures

1) 5	2) 3	3) 56

Similar Figure Word Problems

1) 36.6 ft	3) 14 ft	5) 4 ft
2) 5 in	4) 18 km	

Ratio and Rates Word Problems

1) 210

2) The ratio for both class is equal to 9 to 5.

3) Yes! Both ratios are 1 to 6

4) The price at the Quick Market is a better buy.

5) 640, the rate is 40 per hour.

6) $23.80

Chapter 7: Inequalities

Topics that you'll learn in this chapter:

- ✓ Graphing Single– Variable Inequalities
- ✓ One– Step Inequalities
- ✓ Two– Step Inequalities
- ✓ Multi– Step Inequalities

Graphing Single–Variable Inequalities

Helpful Hints	– Isolate the variable. – Find the value of the inequality on the number line. – For less than or greater than draw open circle on the value of the variable. – If there is an equal sign too, then use filled circle. – Draw a line to the right direction.

✎ *Draw a graph for each inequality.*

1) $-2 > x$

2) $5 \leq -x$

3) $x > 7$

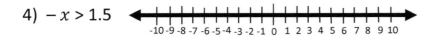

4) $-x > 1.5$

One–Step Inequalities

Helpful *Hints*	– Isolate the variable. – For dividing both sides by negative numbers, flip the direction of the inequality sign.	**Example:** $x + 4 \geq 11$ $x \geq 7$

✎ *Solve each inequality and graph it.*

1) $x + 9 \geq 11$

2) $x - 4 \leq 2$

3) $6x \geq 36$

4) $7 + x < 16$

5) $x + 8 \leq 1$

6) $3x > 12$

7) $3x < 24$

Two–Step Inequalities

Helpful *Hints*	– Isolate the variable.	Example:
	– For dividing both sides by negative numbers, flip the direction of the of the inequality sign.	$2x + 9 \geq 11$
		$2x \geq 2$
	– Simplify using the inverse of addition or subtraction.	$x \geq 1$
	– Simplify further by using the inverse of multiplication or division.	

✎ *Solve each inequality and graph it.*

1) $3x - 4 \leq 5$

2) $2x - 2 \leq 6$

3) $4x - 4 \leq 8$

4) $3x + 6 \geq 12$

5) $6x - 5 \geq 19$

6) $2x - 4 \leq 6$

7) $8x - 4 \leq 4$

8) $6x + 4 \leq 10$

9) $5x + 4 \leq 9$

10) $7x - 4 \leq 3$

11) $4x - 19 < 19$

12) $2x - 3 < 21$

13) $7 + 4x \geq 19$

14) $9 + 4x < 21$

15) $3 + 2x \geq 19$

16) $6 + 4x < 22$

Multi–Step Inequalities

Helpful *Hints*	– Isolate the variable.	**Example:**
	– Simplify using the inverse of addition or subtraction.	$\dfrac{7x + 1}{3} \geq 5$
	– Simplify further by using the inverse of multiplication or division.	$7x + 1 \geq 15$
		$7x \geq 14$
		$x \geq 7$

✎ *Solve each inequality.*

1) $\dfrac{9x}{7} - 7 < 2$

2) $\dfrac{4x + 8}{2} \leq 12$

3) $\dfrac{3x - 8}{7} > 1$

4) $-3\,(x - 7) > 21$

5) $4 + \dfrac{x}{3} < 7$

6) $\dfrac{2x + 6}{4} \leq 10$

Answers of Worksheets – Chapter 7

Graphing Single–Variable Inequalities

1) $-2 > x$

2) $x \leq -5$

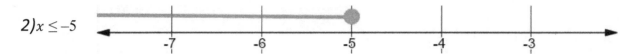

3) $x > 7$

4) $- 1.5 > x$

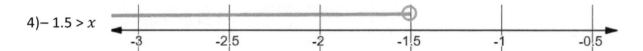

One–Step Inequalities

1)

2)

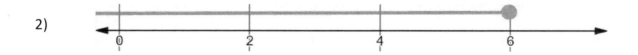

3)

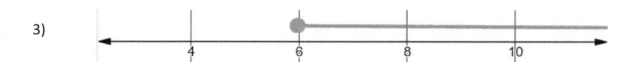

4)

5)

6)

7)

Two–Step inequalities

1) $x \leq 3$

2) $x \leq 4$

3) $x \leq 3$

4) $x \geq 2$

5) $x \geq 4$

6) $x \leq 5$

7) $x \leq 1$

8) $x \leq 1$

9) $x \leq 1$

10) $x \leq 1$

11) $x < 9.5$

12) $x < 12$

13) $x \geq 3$

14) $x < 3$

15) $x \geq 8$

16) $x < 4$

Multi–Step inequalities

1) $x < 7$

2) $x \leq 4$

3) $x > 5$

4) $x < 0$

5) $x < 9$

6) $x \leq 17$

Chapter 8: Exponents and Radicals

Topics that you'll learn in this chapter:

✓ Multiplication Property of Exponents

✓ Division Property of Exponents

✓ Powers of Products and Quotients

✓ Zero and Negative Exponents

✓ Negative Exponents and Negative Bases

✓ Writing Scientific Notation

✓ Square Roots

Multiplication Property of Exponents

Helpful *Hints*	**Exponents rules**	**Example:**
	$x^a \cdot x^b = x^{a+b}$ $x^a/x^b = x^{a-b}$	$(x^2y)^3 = x^6y^3$
	$1/x^b = x^{-b}$ $(x^a)^b = x^{a.b}$	
	$(xy)^a = x^a \cdot y^a$	

✎ *Simplify.*

1) $4^2 \cdot 4^2$

2) $2 \cdot 2^2 \cdot 2^2$

3) $3^2 \cdot 3^2$

4) $3x^3 \cdot x$

5) $12x^4 \cdot 3x$

6) $6x \cdot 2x^2$

7) $5x^4 \cdot 5x^4$

8) $6x^2 \cdot 6x^3y^4$

9) $7x^2y^5 \cdot 9xy^3$

10) $7xy^4 \cdot 4x^3y^3$

11) $(2x^2)^2$

12) $3x^5y^3 \cdot 8x^2y^3$

13) $7x^3 \cdot 10y^3x^5 \cdot 8yx^3$

14) $(x^4)^3$

15) $(2x^2)^4$

16) $(x^2)^3$

17) $(6x)^2$

18) $3x^4y^5 \cdot 7x^2y^3$

Division Property of Exponents

Helpful	$\frac{x^a}{x^b} = x^{a-b}$, $x \neq 0$	**Example:**
Hints		$\frac{x^{12}}{x^5} = x^7$

✎ **Simplify.**

1) $\frac{5^5}{5}$

2) $\frac{3}{3^5}$

3) $\frac{2^2}{2^3}$

4) $\frac{2^4}{2^2}$

5) $\frac{x}{x^3}$

6) $\frac{3x^3}{9x^4}$

7) $\frac{2x^{-5}}{9x^{-2}}$

8) $\frac{21x^8}{7x^3}$

9) $\frac{7x^6}{4x^7}$

10) $\frac{6x^2}{4x^3}$

11) $\frac{5x}{10x^3}$

12) $\frac{3x^3}{2x^5}$

13) $\frac{12x^3}{14x^6}$

14) $\frac{12x^3}{9y^8}$

15) $\frac{25xy^4}{5x^6y^2}$

16) $\frac{2x^4}{7x}$

17) $\frac{16x^2y^8}{4x^3}$

18) $\frac{12x^4}{15x^7y^9}$

19) $\frac{12yx^4}{10yx^8}$

20) $\frac{16x^4y}{9x^8y^2}$

21) $\frac{5x^8}{20x^8}$

Powers of Products and Quotients

Helpful *Hints*	For any nonzero numbers a and b and any integer x, $(ab)^x = a^x \cdot b^x$.	**Example:** $(2x^2 \cdot y^3)^2 =$ $4x^2 \cdot y^6$

✎ *Simplify.*

1) $(2x^3)^4$

2) $(4xy^4)^2$

3) $(5x^4)^2$

4) $(11x^5)^2$

5) $(4x^2y^4)^4$

6) $(2x^4y^4)^3$

7) $(3x^2y^2)^2$

8) $(3x^4y^3)^4$

9) $(2x^6y^8)^2$

10) $(12x \, 3x)^3$

11) $(2x^9 \, x^6)^3$

12) $(5x^{10}y^3)^3$

13) $(4x^3 \, x^2)^2$

14) $(3x^3 \, 5x)^2$

15) $(10x^{11}y^3)^2$

16) $(9x^7 \, y^5)^2$

17) $(4x^4y^6)^5$

18) $(4x^4)^2$

19) $(3x \, 4y^3)^2$

20) $(9x^2y)^3$

21) $(12x^2y^5)^2$

Zero and Negative Exponents

Helpful *Hints*	A negative exponent simply means that the base is on the wrong side of the fraction line, so you need to flip the base to the other side. For instance, "x^{-2}" (pronounced as "ecks to the minus two") just means "x^2" but underneath, as in $\frac{1}{x^2}$	**Example:** $5^{-2} = \frac{1}{25}$

✎ *Evaluate the following expressions.*

1) 8^{-2}

2) 2^{-4}

3) 10^{-2}

4) 5^{-3}

5) 22^{-1}

6) 9^{-1}

7) 3^{-2}

8) 4^{-2}

9) 5^{-2}

10) 35^{-1}

11) 6^{-3}

12) 0^{15}

13) 10^{-9}

14) 3^{-4}

15) 5^{-2}

16) 2^{-3}

17) 3^{-3}

18) 8^{-1}

19) 7^{-3}

20) 6^{-2}

21) $(\frac{2}{3})^{-2}$

22) $(\frac{1}{5})^{-3}$

23) $(\frac{1}{2})^{-8}$

24) $(\frac{2}{5})^{-3}$

25) 10^{-3}

26) 1^{-10}

Negative Exponents and Negative Bases

Helpful	– Make the power positive. A negative exponent is the reciprocal of that number with a positive exponent.	**Example:**
Hints	– The parenthesis is important!	
	-5^{-2} is not the same as $(-5)^{-2}$	$2x^{-3} = \dfrac{2}{x^3}$
	$-5^{-2} = -\dfrac{1}{5^2}$ and $(-5)^{-2} = +\dfrac{1}{5^2}$	

✍ *Simplify.*

1) -6^{-1}

2) $-4x^{-3}$

3) $-\dfrac{5x}{x^{-3}}$

4) $-\dfrac{a^{-3}}{b^{-2}}$

5) $-\dfrac{5}{x^{-3}}$

6) $\dfrac{7b}{-9c^{-4}}$

7) $-\dfrac{5n^{-2}}{10p^{-3}}$

8) $\dfrac{4ab^{-2}}{-3c^{-2}}$

9) $-12x^2y^{-3}$

10) $\left(-\dfrac{1}{3}\right)^{-2}$

11) $\left(-\dfrac{3}{4}\right)^{-2}$

12) $\left(\dfrac{3a}{2c}\right)^{-2}$

13) $\left(-\dfrac{5x}{3yz}\right)^{-3}$

14) $-\dfrac{2x}{a^{-4}}$

Writing Scientific Notation

	– It is used to write very big or very small numbers in decimal form.
Helpful	– In scientific notation all numbers are written in the form of:
Hints	$m \times 10^n$

Decimal notation	Scientific notation
5	5×10^0
−25,000	$−2.5 \times 10^4$
0.5	5×10^{-1}
2,122.456	$2,122456 \times 10^3$

✎ *Write each number in scientific notation.*

1) 91×10^3 8) 0.00023   15) 78

2) 60   9) 56000000   16) 1600

3) 2000000   10) 2000000   17) 1450

4) 0.0000006   11) 78000000   18) 130000

5) 354000   12) 0.0000022   19) 60

6) 0.000325   13) 0.00012   20) 0.113

7) 2.5   14) 0.004   21) 0.02

Square Roots

Helpful	— A square root of x is a number r whose square is: $r^2 = x$	**Example:**
Hints	r is a square root of x.	$\sqrt{4} = 2$

✎ *Find the value each square root.*

1) $\sqrt{1}$

2) $\sqrt{4}$

3) $\sqrt{9}$

4) $\sqrt{25}$

5) $\sqrt{16}$

6) $\sqrt{49}$

7) $\sqrt{36}$

8) $\sqrt{0}$

9) $\sqrt{64}$

10) $\sqrt{81}$

11) $\sqrt{121}$

12) $\sqrt{225}$

13) $\sqrt{144}$

14) $\sqrt{100}$

15) $\sqrt{256}$

16) $\sqrt{289}$

17) $\sqrt{324}$

18) $\sqrt{400}$

19) $\sqrt{900}$

20) $\sqrt{529}$

21) $\sqrt{90}$

Answers of Worksheets – Chapter 8

Multiplication Property of Exponents

1) 4^4

2) 2^5

3) 3^4

4) $3x^4$

5) $36x^5$

6) $12x^3$

7) $25x^8$

8) $36x^5y^4$

9) $63x^3y^8$

10) $28x^4y^7$

11) $4x^4$

12) $24x^7y^6$

13) $560x^{11}y^4$

14) x^{12}

15) $16x^8$

16) x^6

17) $36x^2$

18) $21x^6y^8$

Division Property of Exponents

1) 5^4

2) $\dfrac{1}{3^4}$

3) $\dfrac{1}{2}$

4) 2^2

5) $\dfrac{1}{x^2}$

6) $\dfrac{1}{3x}$

7) $\dfrac{2}{9x^3}$

8) $3x^5$

9) $\dfrac{7}{4x}$

10) $\dfrac{3}{2x}$

11) $\dfrac{1}{2x^2}$

12) $\dfrac{3}{2x^2}$

13) $\dfrac{6}{7x^3}$

14) $\dfrac{4x^3}{3y^8}$

15) $\dfrac{5y^2}{x^5}$

16) $\dfrac{2x^3}{7}$

17) $\dfrac{4y^8}{x}$

18) $\dfrac{4}{5x^3y^9}$

19) $\dfrac{6}{5x^4}$

20) $\dfrac{16}{9x^4y}$

21) $\dfrac{1}{4}$

Powers of Products and Quotients

1) $16x^{12}$

2) $16x^2y^8$

3) $25x^8$

4) $121x^{10}$

5) $256x^8y^{16}$

6) $8x^{12}y^{12}$

7) $9x^4y^4$

8) $81x^{16}y^{12}$

9) $4x^{12}y^{16}$

10) $46,656x^6$

11) $8x^{45}$

12) $125x^{30}y^9$

13) $16x^{10}$

14) $225x^8$

15) $100x^{22}y^6$

16) $81x^{14}y^{10}$

17) $1,024x^{20}y^{30}$

18) $16x^8$

19) $144x^2y^6$ 20) $729x^6y^3$ 21) $144x^4y^{10}$

Zero and Negative Exponents

1) $\frac{1}{64}$

2) $\frac{1}{16}$

3) $\frac{1}{100}$

4) $\frac{1}{125}$

5) $\frac{1}{22}$

6) $\frac{1}{9}$

7) $\frac{1}{9}$

8) $\frac{1}{16}$

9) $\frac{1}{25}$

10) $\frac{1}{35}$

11) $\frac{1}{216}$

12) 0

13) $\frac{1}{1000000000}$

14) $\frac{1}{81}$

15) $\frac{1}{25}$

16) $\frac{1}{8}$

17) $\frac{1}{27}$

18) $\frac{1}{8}$

19) $\frac{1}{343}$

20) $\frac{1}{36}$

21) $\frac{9}{4}$

22) 125

23) 256

24) $\frac{125}{8}$

25) $\frac{1}{1000}$

26) 1

Negative Exponents and Negative Bases

1) $-\frac{1}{6}$

2) $-\frac{4}{x^3}$

3) $-5x^4$

4) $-\frac{b^2}{a^3}$

5) $-5x^3$

6) $-\frac{7bc^4}{9}$

7) $-\frac{p^3}{2n^2}$

8) $-\frac{4ac^2}{3b^2}$

9) $-\frac{12x^2}{y^3}$

10) 9

11) $\frac{16}{9}$

12) $\frac{4c^2}{9a^2}$

13) $-\frac{27y^3z^3}{125x^3}$

14) $-2xa^4$

Writing Scientific Notation

1) 9.1×10^4

2) 6×10^1

3) 2×10^6

4) 6×10^{-7}

5) 3.54×10^5

6) 3.25×10^{-4}

7) 2.5×10^0

8) 2.3×10^{-4}

9) 5.6×10^7

10) 2×10^6

11) 7.8×10^7

12) 2.2×10^{-6}

13) 1.2×10^{-4}

14) 4×10^{-3}

15) 7.8×10^1

16) 1.6×10^3

17) 1.45×10^3

18) 1.3×10^5

19) 6×10^1

20) 1.13×10^{-1}

21) 2×10^{-2}

Square Roots

1) 1

2) 2

3) 3

4) 5

5) 4

6) 7

7) 6

8) 0

9) 8

10) 9

11) 11

12) 15

13) 12

14) 10

15) 16

16) 17

17) 18

18) 20

19) 30

20) 23

21) $3\sqrt{10}$

Chapter 9: Measurements

Topics that you'll learn in this chapter:

- ✓ Inches & Centimeters
- ✓ Metric units
- ✓ Distance Measurement
- ✓ Weight Measurement

Inches and Centimeters

Helpful	1 inch = 2.5 cm	**Example:**
Hints	1 foot = 12 inches	18 inches = 0.4572 m
	1 yard = 3 feet	
	1 yard = 36 inches	
	1 inch = 0.0254 m	

✎ *Convert to the units.*

1inch = 2.5 cm

1) 25 cm = _____ inches

2) 11 inches = _____ cm

3) 1 m = _____ inches

4) 80 inches = _____ m

5) 200 cm = _____ m

6) 5 m = _____ cm

7) 4 feet = _____ inches

8) 10 yards = _____ inches

9) 16 feet = _____ inches

10) 48 inches = _____ Feet

11) 4 inches = _____ cm

12) 12.5 cm = _____ inches

13) 6 feet: _____ inches

14) 10 feet: _____ inches

15) 12 yards: _____ feet

16) 7 yards: _____ feet

Metric Units

Helpful

Hints

1 m = 100 cm

1 cm = 10 mm

1 m = 1000 mm

1 km = 1000 m

Example:

12 cm = 0.12 m

✍ **Convert to the units.**

1) 4 mm = _____ cm

2) 0.6 m = _____ mm

3) 2 m = _____ cm

4) 0.03 km = _____ m

5) 3000 mm = _____ km

6) 5 cm = _____ m

7) 0.03 m = _____ cm

8) 1000 mm = _____ km

9) 600 mm = _____ m

10) 0.77 km = _____ mm

11) 0.08 km = _____ m

12) 0.30 m = _____ cm

13) 400 m = _____ km

14) 5000 cm = _____ km

15) 40 mm = _____ cm

16) 800 m = _____ km

Distance Measurement

Helpful

Hints

1 mile = 5280 ft

1 mile = 1760 yd

1 mile = 1609.34 m

Example:

10 miles = 52800 ft

✍ *Convert to the new units.*

1) 2 mi = _____ ft

2) 21 mi = _____ ft

3) 6 mi = _____ ft

4) 3 mi = _____ yd

5) 72 mi = _____ ft

6) 41 mi = _____ yd

7) 62 mi = _____ yd

8) 39 mi = _____ yd

9) 7 mi = _____ yd

10) 94 mi = _____ yd

11) 87 mi = _____ yd

12) 23 mi = _____ yd

13) 2 mi = _____ m

14) 5 mi = _____ m

15) 6 mi = _____ m

16) 3 mi = _____ m

Weight Measurement

Helpful *Hints*	1 kg = 1000g	Example:
		2000 g = 2 kg

✎ Convert to grams.

1) 0.5 kg = _____ g

2) 3.2 kg = _____ g

3) 8.2 kg = _____ g

4) 9.2 kg = _____ g

5) 35 kg = _____ g

6) 87 kg = _____ g

7) 45 kg = _____ g

8) 15 kg = _____ g

9) 0.32 kg = _____ g

10) 81 kg = _____ g

✎ Convert to kilograms.

11) 200,000 g = _____ kg

12) 30,000 g = _____ kg

13) 800,000 g = _____ kg

14) 20,000 g = _____ kg

15) 40,000 g = _____ kg

16) 500,000 g = _____ kg

Answers of Worksheets – Chapter 9

Inches & Centimeters

1) 25 cm = 9.84 inches

2) 11 inch = 27.94 cm

3) 1 m = 39.37 inches

4) 80 inch = 2.03 m

5) 200 cm = 2 m

6) 5 m = 500 cm

7) 4 feet = 48 inches

8) 10 yards = 360 inches

9) 16 feet = 192 inches

10) 48 inches = 4 Feet

11) 4 inch = 10.16 cm

12) 12.5 cm = 4.92 inches

13) 6 feet: 72 inches

14) 10 feet: 120 inches

15) 12 yards: 36 feet

16) 7 yards: 21 feet

Metric Units

1) 4 mm = 0.4 cm

2) 0.6 m = 600 mm

3) 2 m = 200 cm

4) 0.03 km = 30 m

5) 3000 mm = 0.003 km

6) 5 cm = 0.05 m

7) 0.03 m = 3 cm

8) 1000 mm = 0.001 km

9) 600 mm = 0.6 m

10) 0.77 km = 770,000 mm

11) 0.08 km = 80 m

12) 0.30 m = 30 cm

13) 400 m = 0.4 km

14) 5000 cm = 0.05 km

15) 40 mm = 4 cm

16) 800 m = 0.8 km

Distance Measurement

1) 21 mi = 110880 ft

2) 6 mi = 31680 ft

3) 3 mi = 5280 yd

4) 72 mi = 380160 ft

5) 41 mi = 72160 yd

6) 62 mi = 109120 yd

7) 39 mi = 68640 yd

8) 7 mi = 12320 yd

9) 94 mi = 165440 yd

10) 87 mi = 153120 yd

11) 23 mi = 40480 yd

12) 2 mi = 3218.69 m

13) 5 mi = 8046.72 m

14) 6 mi = 9656.06 m

15) 3 mi = 4828.03 m

Weight Measurement

1) 0.5 kg = 500 g

2) 3.2 kg = 3200 g

3) 8.2 kg = 8200 g

4) 9.2 kg = 9200 g

5) 35 kg = 35000 g

6) 87 kg = 87000 g

7) 45 kg = 45000 g

8) 15 kg = 15000 g

9) 0.32 kg = 320 g

10) 81 kg = 81000 g

11) 200,000 g = 200 kg

12) 30,000 g = 30 kg

13) 800,000 g = 800 kg

14) 20,000 g = 20 kg

15) 40,000 g = 40 kg

16) 500,000 g = 500 kg

Chapter 10: Plane Figures

Topics that you'll learn in this chapter:

- ✓ The Pythagorean Theorem
- ✓ Area of Triangles
- ✓ Perimeter of Polygons
- ✓ Area and Circumference of Circles
- ✓ Area of Squares, Rectangles, and Parallelograms
- ✓ Area of Trapezoids

The Pythagorean Theorem

Helpful

Hints

– In any right triangle:

$a^2 + b^2 = c^2$

Example:

Missing side = 6

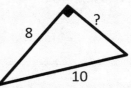

8

?

10

✍ *Do the following lengths form a right triangle?*

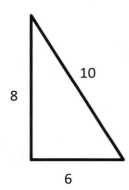

10

8

6

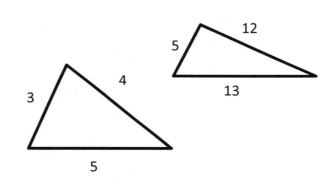

3

4

5

12

5

13

✍ *Find each missing length to the nearest tenth.*

4)

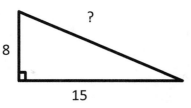

8

?

15

5)

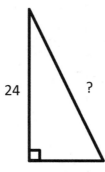

24

?

10

6)

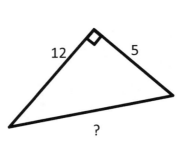

12

5

?

Area of Triangles

Helpful   Area $= \dfrac{1}{2} (base \times height)$

Hints

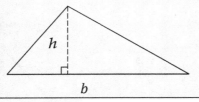

✎ *Find the area of each.*

1)

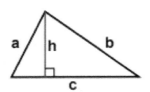

c = 9 mi

h = 3.7 mi

2)

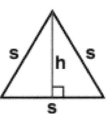

s = 14 m

h = 12.2 m

3)

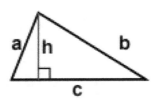

a = 5 m

b = 11 m

c = 14 m

h = 4 m

4)

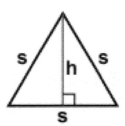

s = 10 m

h = 8.6 m

Perimeter of Polygons

Helpful

Hints

Perimeter of a square = 4s

 s

Perimeter of a rectangle

= 2(l + w)

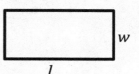 w

l

Perimeter of trapezoid

= a + b + c + d

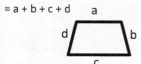
a
d   b
c

Perimeter of Pentagon = 6a

a

Perimeter of a parallelogram = 2(l + w)

l

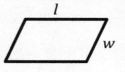 w

Example:

P = 18

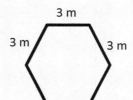
3 m

3 m   3 m

🖉 *Find the perimeter of each shape.*

1)
5 m
5 m   5 m

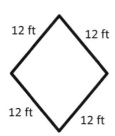

2)
15 mm

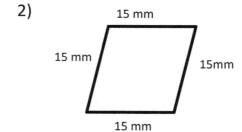

15 mm   15mm

15 mm

3)
12 ft   12 ft

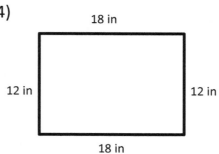

12 ft   12 ft

4)
18 in

12 in   12 in

18 in

Area and Circumference of Circles

Helpful

Hints

Area = πr²

Circumference = 2πr

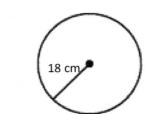

Example:

If the radius of a circle is 3, then:

Area = 28.27

Circumference = 18.85

✐*Find the area and circumference of each.* (π = 3.14)

1)

4 in

2)

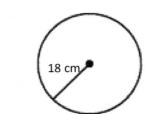

18 cm

3)

5 m

4)

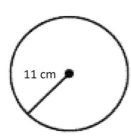

11 cm

5)

8 km

6)

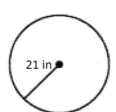

21 in

Area of Squares, Rectangles, and Parallelograms

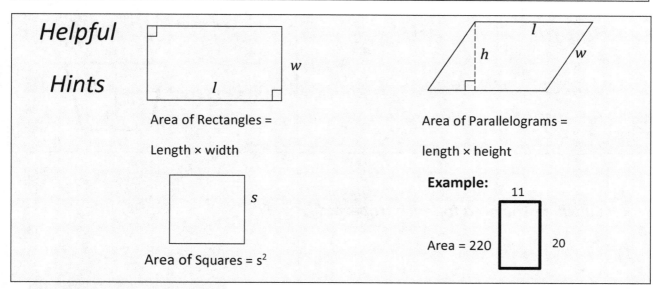

Helpful Hints

Area of Rectangles = Length × width

Area of Squares = s^2

Area of Parallelograms = length × height

Example:

11

Area = 220   20

✏️ *Find the area of each.*

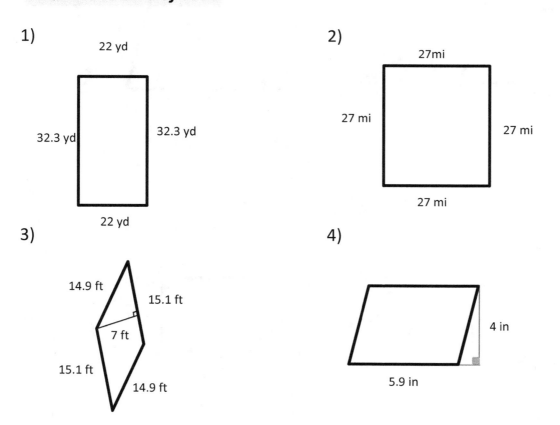

1)

22 yd

32.3 yd   32.3 yd

22 yd

2)

27 mi

27 mi   27 mi

27 mi

3)

14.9 ft

15.1 ft

7 ft

15.1 ft

14.9 ft

4)

4 in

5.9 in

Area of Trapezoids

Helpful $A = \frac{1}{2}h(b_1 + b_2)$ **Example:**

Hints

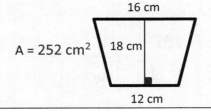

A = 252 cm²

16 cm

18 cm

12 cm

✎ *Calculate the area for each trapezoid.*

1)

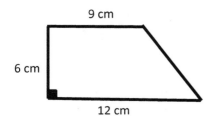

9 cm

6 cm

12 cm

2)

14 m

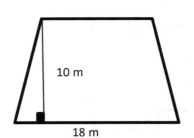

10 m

18 m

3)

22 mi

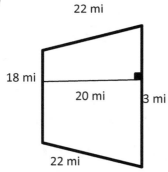

18 mi

20 mi

3 mi

22 mi

4)

8.6 nm

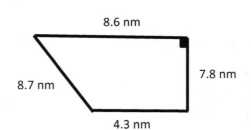

8.7 nm

7.8 nm

4.3 nm

Answers of Worksheets – Chapter 10

The Pythagorean Theorem

1) yes

2) yes

3) yes

4) 17

5) 26

6) 13

Area of Triangles

1) 16.65 mi^2

2) 85.4 m^2

3) 28 m^2

4) 43 m^2

Perimeter of Polygons

1) 30 m

2) 60 mm

3) 48 ft

4) 60 in

Area and Circumference of Circles

1) Area: 50.24 in^2, Circumference: 25.12 in

2) Area: 1,017.36 cm^2, Circumference: 113.04 cm

3) Area: 78.5m^2, Circumference: 31.4 m

4) Area: 379.94 cm^2, Circumference: 69.08 cm

5) Area: 200.96 km^2, Circumference: 50.2 km

6) Area: 1,384.74 km^2, Circumference: 131.88 km

Area of Squares, Rectangles, and Parallelograms

1) 710.6 yd^2

2) 729 mi^2

3) 105.7 ft^2

4) 23.6 in^2

Area of Trapezoids

1) 63 cm^2

2) 160 m^2

3) 410 mi^2

4) 50.31 nm^2

Chapter 11: Solid Figures

Topics that you'll learn in this chapter:

- ✓ Volume of Cubes
- ✓ Volume of Rectangle Prisms
- ✓ Surface Area of Cubes
- ✓ Surface Area of Rectangle Prisms

Volume of Cubes

Helpful	– Volume is the measure of the amount of space inside of a solid figure, like a cube, ball, cylinder or pyramid.
Hints	– Volume of a cube = (one side)^3
	– Volume of a rectangle prism: Length × Width × Height

✎ *Find the volume of each.*

1)

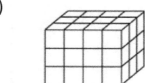

2)

3)

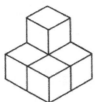

4)

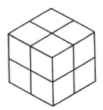

5)

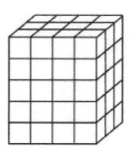

6)

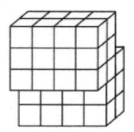

Volume of Rectangle Prisms

Helpful

Hints

Volume of rectangle prism

length × width × height

Example:

$10 \times 5 \times 8 = 400m^3$

10 m

8 m

5 m

✎ *Find the volume of each of the rectangular prisms.*

1)

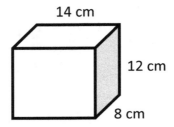

14 cm

12 cm

8 cm

2)

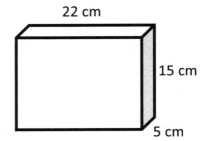

22 cm

15 cm

5 cm

3)

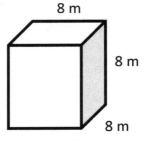

8 m

8 m

8 m

4)

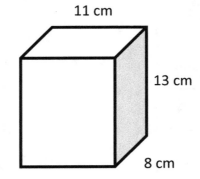

11 cm

13 cm

8 cm

Surface Area of Cubes

Helpful

Hints

Surface Area of a cube =

6 × (one side of the cube)^2

Example:

6 × 4^2 = 96m^2

4 m

4 m

4 m

✎ *Find the surface of each cube.*

1)

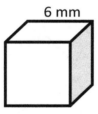

6 mm

2)

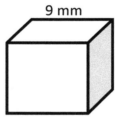

9 mm

3)

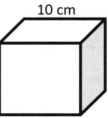

10 cm

4)

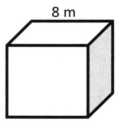

8 m

5)

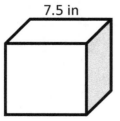

7.5 in

6)

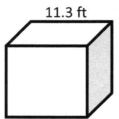

11.3 ft

Surface Area of a Rectangle Prism

Helpful

Hints

Surface Area of a Rectangle Prism Formula:

SA =2 [(width × length) + (height × length) + width × height)]

🖎 *Find the surface of each prism.*

1)

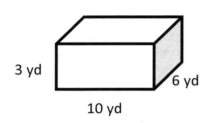

3 yd

6 yd

10 yd

2)

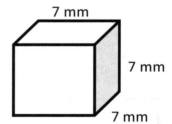

7 mm

7 mm

7 mm

3)

8 in

13.2 in

6.7 in

4)

17 cm

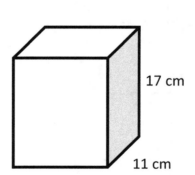

17 cm

11 cm

Volume of a Cylinder

Helpful

Hints

Volume of Cylinder Formula = π(radius)² × height

π = 3.14

✎ **Find the volume of each cylinder.** (π = 3.14)

1)

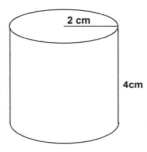

2 cm

4cm

2)

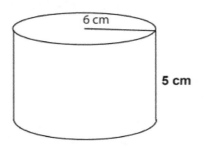

6 cm

5 cm

3)

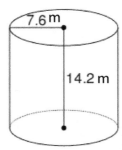

7.6 m

14.2 m

4)

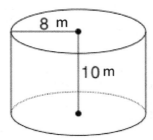

8 m

10 m

Surface Area of a Cylinder

Helpful		Example:
	Surface area of a cylinder	Surface area
Hints	$SA = 2\pi r^2 + 2\pi rh$	= 1727

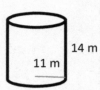

14 m

11 m

✎ **Find the surface of each cylinder.** ($\pi = 3.14$)

1)

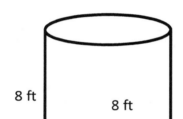

8 ft

8 ft

2)

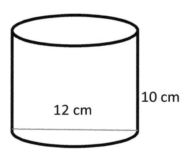

10 cm

12 cm

3)

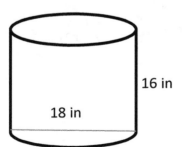

16 in

18 in

4)

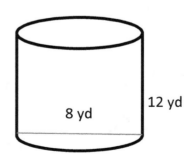

12 yd

8 yd

Answers of Worksheets – Chapter 11

Volumes of Cubes

1) 8

2) 4

3) 5

4) 36

5) 60

6) 44

Volume of Rectangle Prisms

1) 1344 cm^3

2) 1650 cm^3

3) 512 m^3

4) 1144 cm^3

Surface Area of a Cube

1) 216 mm^2

2) 486 mm^2

3) 600 cm^2

4) 384 m^2

5) 337.5 in^2

6) 766.14 ft^2

Surface Area of a Prism

1) 216 yd^2

2) 294 mm^2

3) 495.28 in^2

4) 1326 cm^2

Volume of a Cylinder

1) 50.24 cm^3

2) 565.2 cm^3

3) 2,575.403 m^3

4) 2009.6 m^3

Surface Area of a Cylinder

1) 301.44 ft^2

2) 602.88 cm^2

3) 1413 in^2

4) 401.92 yd^2

Chapter 12: Statistics

Topics that you'll learn in this chapter:

✓ Mean, Median, Mode, and Range of the Given Data

✓ The Pie Graph or Circle Graph

✓ Probability

Mean, Median, Mode, and Range of the Given Data

Helpful	- Mean: $\dfrac{\text{sum of the data}}{\text{of data entires}}$	**Example:**
Hints	- Mode: value in the list that appears most often - Range: largest value – smallest value	22, 16, 12, 9, 7, 6, 4, 6 Mean = 10.25 Mod = 6 Range = 18

✎ *Find Mean, Median, Mode, and Range of the Given Data.*

1) 7, 2, 5, 1, 1, 2

2) 2, 2, 2, 3, 6, 3, 7, 4

3) 9, 4, 3, 1, 7, 9, 4, 6, 4

4) 8, 4, 2, 4, 3, 2, 4, 5

5) 8, 5, 7, 5, 7, 9, 8

6) 5, 1, 4, 4, 9, 2, 9, 2, 5, 1

7) 4, 1, 5, 9, 7, 7, 5, 4, 3, 5

8) 7, 5, 4, 9, 6, 7, 7, 5, 2

9) 2, 5, 5, 6, 2, 4, 7, 6, 4, 9

10) 10, 5, 2, 5, 4, 5, 8, 10

11) 5, 1, 5, 2, 2

12) 2, 3, 5, 9, 6

The Pie Graph or Circle Graph

Helpful	A Pie Chart is a circle chart divided into sectors, each sector represents the relative size of each value.
Hints	

The circle graph below shows all Jason's expenses for last month. Jason spent $300 on his bills last month.

Answer following questions based on the Pie graph.

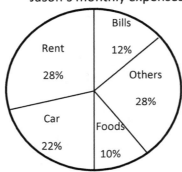

Jason's monthly expenses

1- How much did Jason spend on his car last month?

2- How much did Jason spend for foods last month?

3- How much did Jason spend on his rent last month?

4- What fraction is Jason's expenses for his bills and Car out of his total expenses last month?

5- How much was Jason's senses last month?

Probability of Simple Events

Helpful *Hints*	- Probability is the likelihood of something happening in the future. It is expressed as a number between zero (can never happen) to 1 (will always happen). - Probability can be expressed as a fraction, a decimal, or a percent.	**Example:** Probability of a flipped coins turns up 'heads' Is $0.5 = \dfrac{1}{2}$

✎ *Solve.*

1) A number is chosen at random from 1 to 10. Find the probability of selecting a 4 or smaller.

2) There are 135 blue balls and 15 red balls in a basket. What is the probability of randomly choosing a red ball from the basket?

3) A number is chosen at random from 1 to 10. Find the probability of selecting of 4 and factors of 6.

4) What is the probability of choosing a Hearts in a deck of cards? (A deck of cards contains 52 cards)

5) A number is chosen at random from 1 to 50. Find the probability of selecting prime numbers.

6) A number is chosen at random from 1 to 25. Find the probability of not selecting a composite number.

Answers of Worksheets – Chapter 12

Mean, Median, Mode, and Range of the Given Data

1) mean: 3, median: 2, mode: 1, 2, range: 6
2) mean: 3.625, median: 3, mode: 2, range: 5
3) mean: 5.22, median: 4, mode: 4, range: 8
4) mean: 4, median: 4, mode: 4, range: 6
5) mean: 7, median: 7, mode: 5, 7, 8, range: 4
6) mean: 4.2, median: 4, mode: 1,2,4,5,9, range: 8
7) mean: 5, median: 5, mode: 5, range: 8
8) mean: 5.78, median: 6, mode: 7, range: 7
9) mean: 5, median: 5, mode: 2, 4, 5, 6, range: 7
10) mean: 6.125, median: 5, mode: 5, range: 8
11) mean: 3, median: 2, mode: 2, 5, range: 4
12) mean: 5, median: 5, mode: none, range: 7

The Pie Graph or Circle Graph

1) $550
2) $250
3) $700

4) $\frac{17}{50}$
5) $2500

Probability of simple events

1) $\frac{2}{5}$

2) $\frac{1}{10}$

3) $\frac{1}{5}$

4) $\frac{1}{4}$

5) $\frac{3}{10}$

6) $\frac{2}{5}$

ATI TEAS 6 Test Review

The ATI TEAS (Test of Essential Academic Skills), known as TEAS, is an admissions test for nursing schools, and is designed to assess a student's preparedness entering the health science fields. The last edition (the sixth edition) of the test, called the ATI TEAS 6 Test, was published by ATI Testing on August 31, 2016.

The ATI TEAS 6 Test consists of four multiple-choice sections:

- ✓ **Reading:** 53 Questions – 64 Minutes
- ✓ **Mathematics:** 36 Questions – 54 Minutes
- ✓ **Science:** 53 Questions – 63 Minutes
- ✓ **English and Language Usage:** 28 Questions – 28 Minutes

The Math portion will consist of around 36 multiple-choice questions that address
The Math section of the test covers two main topics: Number and Algebra; Measurement and Data.

Students will be allowed to use a four-function calculator during the Math section of the ATI TEAS test. A calculator will be included in the online version and students will be issued one at the testing center during a paper and pencil test.

In this section, there are two complete ATI TEAS 6 Mathematics Tests. Take these tests to see what score you'll be able to receive on a real TEAS test.

Good luck!

ATI TEAS 6 Mathematics Practice Tests

Time to Test

Time to refine your skill with a practice examination

Take practice ATI TEAS 6 Math Tests to simulate the test day experience. After you've finished, score your tests using the answer keys.

Before You Start

- You'll need a pencil, a timer, and a four-function calculator to take the test.

- After you've finished the test, review the answer key to see where you went wrong.

- Use the answer sheet provided to record your answers. (You can cut it out or photocopy it)

- You will receive 1 point for every correct answer. There is no penalty for wrong answers.

Good Luck!

ATI TEAS 6 Mathematics Practice Tests Answer Sheets

Remove (or photocopy) these two answer sheets and use them to complete the practice tests.

ATI TEAS 6 Mathematics Practice Test 1 Answer Sheet		
1 Ⓐ Ⓑ Ⓒ Ⓓ	13 Ⓐ Ⓑ Ⓒ Ⓓ	25 Ⓐ Ⓑ Ⓒ Ⓓ
2 Ⓐ Ⓑ Ⓒ Ⓓ	14 Ⓐ Ⓑ Ⓒ Ⓓ	26 Ⓐ Ⓑ Ⓒ Ⓓ
3 Ⓐ Ⓑ Ⓒ Ⓓ	15 Ⓐ Ⓑ Ⓒ Ⓓ	27 Ⓐ Ⓑ Ⓒ Ⓓ
4 Ⓐ Ⓑ Ⓒ Ⓓ	16 Ⓐ Ⓑ Ⓒ Ⓓ	28 Ⓐ Ⓑ Ⓒ Ⓓ
5 Ⓐ Ⓑ Ⓒ Ⓓ	17 Ⓐ Ⓑ Ⓒ Ⓓ	29 Ⓐ Ⓑ Ⓒ Ⓓ
6 Ⓐ Ⓑ Ⓒ Ⓓ	18 Ⓐ Ⓑ Ⓒ Ⓓ	30 Ⓐ Ⓑ Ⓒ Ⓓ
7 Ⓐ Ⓑ Ⓒ Ⓓ	19 Ⓐ Ⓑ Ⓒ Ⓓ	31 Ⓐ Ⓑ Ⓒ Ⓓ
8 Ⓐ Ⓑ Ⓒ Ⓓ	20 Ⓐ Ⓑ Ⓒ Ⓓ	32 Ⓐ Ⓑ Ⓒ Ⓓ
9 Ⓐ Ⓑ Ⓒ Ⓓ	21 Ⓐ Ⓑ Ⓒ Ⓓ	33 Ⓐ Ⓑ Ⓒ Ⓓ
10 Ⓐ Ⓑ Ⓒ Ⓓ	22 Ⓐ Ⓑ Ⓒ Ⓓ	34 Ⓐ Ⓑ Ⓒ Ⓓ
11 Ⓐ Ⓑ Ⓒ Ⓓ	23 Ⓐ Ⓑ Ⓒ Ⓓ	35 Ⓐ Ⓑ Ⓒ Ⓓ
12 Ⓐ Ⓑ Ⓒ Ⓓ	24 Ⓐ Ⓑ Ⓒ Ⓓ	36 Ⓐ Ⓑ Ⓒ Ⓓ

ATI TEAS 6 Mathematics Practice Test 2 Answer Sheet		
1 (A) (B) (C) (D)	13 (A) (B) (C) (D)	25 (A) (B) (C) (D)
2 (A) (B) (C) (D)	14 (A) (B) (C) (D)	26 (A) (B) (C) (D)
3 (A) (B) (C) (D)	15 (A) (B) (C) (D)	27 (A) (B) (C) (D)
4 (A) (B) (C) (D)	16 (A) (B) (C) (D)	28 (A) (B) (C) (D)
5 (A) (B) (C) (D)	17 (A) (B) (C) (D)	29 (A) (B) (C) (D)
6 (A) (B) (C) (D)	18 (A) (B) (C) (D)	30 (A) (B) (C) (D)
7 (A) (B) (C) (D)	19 (A) (B) (C) (D)	31 (A) (B) (C) (D)
8 (A) (B) (C) (D)	20 (A) (B) (C) (D)	32 (A) (B) (C) (D)
9 (A) (B) (C) (D)	21 (A) (B) (C) (D)	33 (A) (B) (C) (D)
10 (A) (B) (C) (D)	22 (A) (B) (C) (D)	34 (A) (B) (C) (D)
11 (A) (B) (C) (D)	23 (A) (B) (C) (D)	35 (A) (B) (C) (D)
12 (A) (B) (C) (D)	24 (A) (B) (C) (D)	36 (A) (B) (C) (D)

ATI TEAS 6 Mathematics Practice Test 1

- ○ **36 questions**

- ○ **Total time for this section:** 54 Minutes

- ○ **Calculator is allowed at the test.**

1) If $(4.2 + 4.3 + 4.5)\, x = x$, then what is the value of x?

   A. 0

   B. $\frac{1}{10}$

   C. 1

   D. 10

2) The sum of two numbers is x. If one of the numbers is 9, then two times the other number would be?

   A. $2x$

   B. $2 + x \times 2$

   C. $2(x + 9)$

   D. $2(x - 9)$

3) A circle has a diameter of 8 inches. What is its approximate circumference?

   A. 6.28

   B. 25.12

   C. 34.85

   D. 35.12

4) If $x = \frac{5}{7}$ then $\frac{1}{x} = ?$

   A. $\frac{7}{5}$

   B. $\frac{5}{7}$

   C. 5

   D. 7

5) If $6.5 < x \le 9.0$, then x cannot be equal to:

   A. 6.5

   B. 9

   C. 7.2

   D. 10.5

6) Which of the following is the product of $1\frac{1}{2}$ and $4\frac{3}{5}$?

   A. $9\frac{6}{10}$

   B. $6\frac{9}{10}$

   C. $10\frac{6}{9}$

   D. $4\frac{2}{3}$

7) How many ¼ pound paperback books together weigh 30 pounds?

    A. 80

    B. 95

    C. 105

    D. 120

8) $\frac{7}{25}$ is equals to:

    A. 0.3

    B. 2.8

    C. 0.03

    D. 0.28

9) In the simplest form, $\frac{18}{24}$ is

    A. $\frac{2}{3}$

    B. $\frac{3}{2}$

    C. $\frac{4}{3}$

    D. $\frac{3}{4}$

10) The sum of 8 numbers is greater than 240 and less than 320. Which of the following could be the average (arithmetic mean) of the numbers?

A. 30

B. 35

C. 40

D. 45

11) If $x = 6$, then $\dfrac{6^5}{x} =$

A. 30

B. 7,776

C. 1,296

D. 96

12) The perimeter of a rectangular yard is 60 meters. What is its length if its width is twice its length?

A. 10 meters

B. 18 meters

C. 20 meters

D. 24 meters

13) A swimming pool holds 2,000 cubic feet of water. The swimming pool is 25 feet long and 10 feet wide. How deep is the swimming pool?

A. 6

B. 8

C. 10

D. 12

14) Chris is 9 miles ahead of Joe running at 5.5 miles per hour and Joe is running at the speed of 7 miles per hour. How long does it take Joe to catch Chris?

A. 3 hours

B. 4 hours

C. 6 hours

D. 8 hours

15) $\dfrac{(15 \text{ feet} + 7 \text{ yards})}{4} = $ ___

A. 9 feet

B. 7 feet

C. 28 feet

D. 4 feet

16) A bread recipe calls for $3\frac{1}{3}$ cups of flour. If you only have $2\frac{5}{6}$ cups, how much more flour is needed?

   A. 1

   B. $\frac{1}{2}$

   C. 2

   D. $\frac{7}{6}$

17) The equation of a line is given as : $y = 5x - 3$. Which of the following points does not lie on the line?

   A. (1, 2)

   B. (−2, −13)

   C. (3, 18)

   D. (2, 7)

18) If $a = 8$, what is the value of b in the following equation?

$$b = \frac{a^2}{4} + c$$

   A. $8 + c$

   B. $2 + c$

   C. $16 + c$

   D. $24 + c$

19) If a circle has a radius of 29 feet, what's the closest approximation of its circumference?

   A. 87

   B. 209

   C. 182

   D. 58

20) 8 feet, 10 inches + 5 feet, 12 inches equals to how many inches?

   A. 178 inches

   B. 188 inches

   C. 182 inches

   D. 200 inches

21) The distance between cities A and B is approximately 2,600 miles. If you drive an average of 68 miles per hour, how many hours will it take you to drive from city A to city B?

   A. approximately 41 hours

   B. approximately 38 hours

   C. approximately 29 hours

   D. approximately 27 hours

22) If two angles in a triangle measure 53 degrees and 45 degrees, what is the value of the third angle?

A. 8 degrees

B. 42 degrees

C. 98 degrees

D. 82 degrees

23) What is the equivalent temperature of 104°F in Celsius?

$$C = \frac{5}{9}(F - 32)$$

A. 32

B. 40

C. 48

D. 52

Questions 24 and 25 are based on following Pie Chart.

The following pie chart shows the time Jason spent to work on his homework last week.

The total time Jason spent on his homework last week was 20 hours.

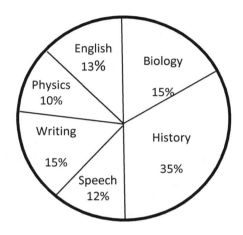

24) How much time did Jason spend on History last week?

    A. 3.5 hours

    B. 5 hours

    C. 6.5 hours

    D. 7 hours.

25) What many hours did Jason spend doing the Writing and Physics?

    A. 3 hours

    B. 5 hours

    C. 8 hours

    D. 10 hours

26) Julie gives 8 pieces of candy to each of her friends. If Julie gives all her candy away, which amount of candy could have been the amount she distributed?

A. 187

B. 216

C. 243

D. 223

27) With an 22% discount, Ella was able to save $20.42 on a dress. What was the original price of the dress?

A. $88.92

B. $90.82

C. $92.82

D. $93.92

28) What is the sum of $\frac{1}{3} + \frac{2}{5} + \frac{1}{2}$?

A. 0.9

B. 1

C. 1.23

D. $1\frac{7}{30}$

0.ABC   0.0D

29) The letters represent two decimals listed above. One of the decimals is equivalent to $\frac{1}{8}$ and the other is equivalent to $\frac{1}{20}$. What is the product of C and D?

(A)   0

(B)   5

(C)   25

(D)   20

30) The marked price of a computer is D dollar. Its price decreased by 20% in January and later increased by 10 % in February. What is the final price of the computer in D dollar?

(A)   0.80 D

(B)   0.88 D

(C)   0.90 D

(D)   1.20 D

31) If three times a number added to 6 equals to 30, what is the number?

(A) 2

(B) 4

(C) 6

(D) 8

32) A bank is offering 3.5% simple interest on a savings account. If you deposit $12,000, how much interest will you earn in two years?

A. $420

B. $840

C. $4,200

D. $8,400

33) What is the value of x in the following equation?

$$\frac{2}{3}x + \frac{1}{6} = \frac{1}{3}$$

A.   6

B. $\frac{1}{2}$

C. $\frac{1}{3}$

D. $\frac{1}{4}$

34) A football team had $20,000 to spend on supplies. The team spent $14,000 on new balls. New sport shoes cost $120 each. Which of the following inequalities represent how many new shoes the team can purchase.

A. $120x + 14,000 \leq 20,000$

B. $120x + 14,000 \geq 20,000$

C. $14,000x + 120 \leq 20,000$

D. $14,000x + 12,0 \geq 20,000$

35) Two dice are thrown simultaneously, what is the probability of getting a sum of 6 or 9?

A. $\dfrac{1}{3}$

B. $\dfrac{1}{4}$

C. $\dfrac{1}{6}$

D. $\dfrac{1}{12}$

36) Which of the following graphs represents the compound inequality $-2 \le 2x - 4 < 8$?

A.

B.

C.

D.

ATI TEAS 6 Mathematics
Practice Test 2

- **36 questions**
- **Total time for this section:** 54 Minutes
- **Calculator is allowed for the test.**

1) If $x + y = 12$, what is the value of $8x + 8y$?

   A. 192

   B. 48

   C. 104

   D. 96

2) If $6 + x \geq 18$, then $x \geq$?

   A. 3

   B. 6

   C. 12

   D. $18x$

3) What is the sum of $\frac{7}{12} + \frac{4}{3} + \frac{2}{6}$?

   A. 2.25

   B. 2

   C. $2\frac{2}{3}$

   D. 1

4) If a rectangle is 30 feet by 45 feet, what is its area?

   A. 1,350

   B. 870

   C. 1,000

   D. 1,250

5) If x is 25% percent of 250, what is x?

   A. 35

   B. 95.5

   C. 62.5

   D. 150

6) The width of a garden is 5.48 yards. How many meters is the width of that garden?

   A. 5.01

   B. 301.1 m

   C. 57.4 m

   D. 195.1 m

7) The oven temperature reaches 55° C. what's is the temperature in degree Fahrenheit?

$$C = \frac{5}{9} (F - 32)$$

A. 55° F

B. 131° F

C. 77° F

D. 95° F

8) How many meters is 27,356 centimeters?

A. 27.356 m

B. 2.735600 m

C. 273.5600 m

D. 2,735.600 m

9) In five successive hours, a car travels 40 km, 45 km, 50 km, 35 km and 55 km. In the next five hours, it travels with an average speed of 50 km per hour. Find the total distance the car traveled in 10 hours.

(A)   425 km

(B)   450 km

(C)   475 km

(D)   500 km

10) Find the mean of 112, 420, 322, 176, 390, and 235.

    A. 250.5

    B. 275.833 …

    C. 277.5

    D. 292

11) Solve the proportion. $\frac{2.5}{3.2} = \frac{x}{5.6}$

    A. 3.457

    B. 1.745

    C. 2.547

    D. 4.375

12) The equation of a line is given as: $y = 5x - 3$. Which of the following points does not lie on the line?

    A. (1, 2)

    B. (−2, −13)

    C. (3, 18)

    D. (2, 7)

13) If two angles in a triangle measure 50 degrees and 42 degrees, what is the value of the third angle?

A. 88 Degrees

B. 42 Degrees

C. 92 Degrees

D. 112 Degrees

14) Ella (E) is 4 years older than her friend Ava (A) who is 3 years younger than her sister Sofia (S). If E, A and S denote their ages, which one of the following represents the given information?

A. $\begin{cases} E = A + 4 \\ S = A - 3 \end{cases}$

B. $\begin{cases} E = A + 4 \\ A = S + 3 \end{cases}$

C. $\begin{cases} A = E + 4 \\ S = A - 3 \end{cases}$

D. $\begin{cases} E = A + 4 \\ A = S - 3 \end{cases}$

15) Five years ago, Amy was three times as old as Mike was. If Mike is 10 years old now, how old is Amy?

    (A) 4

    (B) 8

    (C) 15

    (D) 20

16) A number is chosen at random from 1 to 25. Find the probability of not selecting a composite number.

    A. $\dfrac{9}{25}$

    B. 25

    C. $\dfrac{2}{5}$

    D. 1

17) Last Friday Jacob had $32.52. Over the weekend he received some money for cleaning the attic. He now has $44. How much money did he receive?

    A. $76.52

    B. $11.48

    C. $32.08

    D. $12.58

18) Simplify $\dfrac{\dfrac{1}{2} - \dfrac{x+5}{4}}{\dfrac{x^2}{2} - \dfrac{5}{2}}$

A. $\dfrac{3-x}{x^2-10}$

B. $\dfrac{3-x}{2x^2-10}$

C. $\dfrac{3+x}{x^2-10}$

D. $\dfrac{-3-x}{2x^2-10}$

19) Two-kilograms apple and three-kilograms orange cost \$26.4. If one-kilogram apple costs \$4.2 how much does one-kilogram orange cost?

(A) \$9

(B) \$6

(C) \$5.5

(D) \$5

20) The average weight of 18 girls in a class is 60 kg and the average weight of 32 boys in the same class is 62 kg. What is the average weight of all the 50 students in that class?

(A) 61.28

(B) 61.68

(C) 61.90

(D) 62.20

21) What is the value of x in this equation? $6(x + 4) = 72$

    (A)   4

    (B)   6

    (C)   8

    (D)   10

22) A circle has a diameter of 3.8 inches. What is its approximate circumference?

    A. 11

    B. 12

    C. 13

    D. 14

23) In a certain bookshelf of a library, there are 35 biology books, 95 history books, and 80 language books. What is the ratio of the number of biology books to the total number of books in this bookshelf?

    (A) $\frac{1}{4}$

    (B) $\frac{1}{6}$

    (C) $\frac{2}{7}$

    (D) $\frac{3}{8}$

24) If $x = 7$ what's the value of $6x^2 + 5x - 13$?

   A. 64

   B. 316

   C. 416

   D. 293

25) The circumference of a circle is 30 cm. what is the approximate radius of the circle?

   A. 2.4 cm

   B. 4.8 cm

   C. 8.0 cm

   D. 9.5 cm

26) At a Zoo, the ratio of lions to tigers is 5 to 3. Which of the following could NOT be the total number of lions and tigers in the zoo?

   (A)   64

   (B)   80

   (C)   98

   (D)   104

27) I've got 34 quarts of milk and my family drinks 2 gallons of milk per week. How many weeks will that last us?

A. 2 Weeks

B. 2.5 Weeks

C. 3.25 Weeks

D. 4.25 Weeks

28) If a inches of rain falls in one minute, how many inches will fall in b hours?

A. $60\dfrac{a}{b}$

B. $60a$

C. $60b$

D. $60ab$

Questions 29 and 30 are based on following chart.

The following pie chart shows the expenses of Mr. Jackson's family in December.

The total expenses in December was $5,200.

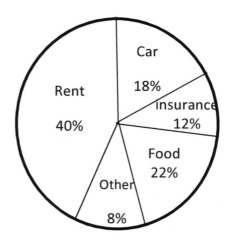

29) What percent of the expenses goes for car and rent combined?

A. 40 %

B. 48 %

C. 58 %

D. 66 %

30) How much did Mr. Jackson's family spend on insurance?

A. $416

B. $520

C. $624

D. $1,040

31) In a bundle of 90 pencils, 43 are red and the rest are blue. What percent of the bundle is composed of blue pencils?

A. 53%

B. 50%

C. 54%

D. 52%

32) A football team won exactly 80% of the games it played during last session. Which of the following could be the total number of games the team played last season?

A. 45

B. 35

C. 32

D. 12

33) A card is drawn at random from a standard 52–card deck, what is the probability that the card is of Hearts? (The deck includes 13 of each suit clubs, diamonds, hearts, and spades)

A. $\frac{1}{3}$

B. $\frac{1}{4}$

C. $\frac{1}{6}$

D. $\frac{1}{52}$

34) What is the value of the expression $5(x - 2y) + (2 - x)^2$ when

$x = 3$ and $= -2$?

A. 4

B. 20

C. 36

D. 50

35) If a gas tank can hold 25 gallons, how many gallons does it contain when it is $\frac{2}{5}$ full?

(A)   50

(B)   125

(C)   62.5

(D)   10

36) 6 liters of water are poured into an aquarium that's 15 cm long, 5 cm wide, and 60 cm high. How many cm will the water level in the aquarium rise due to this added water? (1 liter of water = 1,000 cm^3)

(A)   80

(B)   40

(C)   20

(D)   8

ATI TEAS Mathematics

Practice Tests Answers

and Explanations

ATI TEAS Practice Test 1
Answers

1-	A	19-	C
2-	D	20-	A
3-	B	21-	B
4-	A	22-	D
5-	A	23-	B
6-	B	24-	D
7-	D	25-	B
8-	D	26-	B
9-	C	27-	C
10-	B	28-	C
11-	C	29-	C
12-	A	30-	B
13-	B	31-	D
14-	C	32-	B
15-	A	33-	D
16-	B	34-	A
17-	C	35-	B
18-	C	36-	D

ATI TEAS Practice Test 2
Answers

1-	D	19-	B
2-	C	20-	A
3-	A	21-	C
4-	A	22-	B
5-	C	23-	B
6-	A	24-	B
7-	B	25-	B
8-	C	26-	C
9-	C	27-	D
10-	B	28-	D
11-	D	29-	C
12-	C	30-	C
13-	A	31-	D
14-	D	32-	B
15-	D	33-	B
16-	C	34-	C
17-	B	35-	D
18-	D	36-	A

"Effortless Math" Publications

Effortless Math authors' team strives to prepare and publish the best quality Mathematics learning resources to make learning Math easier for all. We hope that our publications help you or your student Math in an effective way.

We all in Effortless Math wish you good luck and successful studies!

Effortless Math Authors

www.EffortlessMath.com

... So Much More Online!

✓ FREE Math lessons

✓ More Math learning books!

✓ Mathematics Worksheets

✓ Online Math Tutors

Need a PDF version of this book?

Visit www.EffortlessMath.com

Or send email to: info@EffortlessMath.com

9 781970 036077